T0140150

Lecture Notes in Networks and Systems

Volume 310

The series "Lecture Notes in Networks and Systems" publishes the latest developments in Networks and Systems—quickly, informally and with high quality. Original research reported in proceedings and post-proceedings represents the core of LNNS.

Volumes published in LNNS embrace all aspects and subfields of, as well as new challenges in, Networks and Systems.

The series contains proceedings and edited volumes in systems and networks, spanning the areas of Cyber-Physical Systems, Autonomous Systems, Sensor Networks, Control Systems, Energy Systems, Automotive Systems, Biological Systems, Vehicular Networking and Connected Vehicles, Aerospace Systems, Automation, Manufacturing, Smart Grids, Nonlinear Systems, Power Systems, Robotics, Social Systems, Economic Systems and other. Of particular value to both the contributors and the readership are the short publication timeframe and the world-wide distribution and exposure which enable both a wide and rapid dissemination of research output.

The series covers the theory, applications, and perspectives on the state of the art and future developments relevant to systems and networks, decision making, control, complex processes and related areas, as embedded in the fields of interdisciplinary and applied sciences, engineering, computer science, physics, economics, social, and life sciences, as well as the paradigms and methodologies behind them.

Indexed by SCOPUS, INSPEC, WTI Frankfurt eG, zbMATH, SCImago.

All books published in the series are submitted for consideration in Web of Science.

More information about this series at http://www.springer.com/series/15179

Kim-Kwang Raymond Choo ·
Tommy Morris · Gilbert Peterson ·
Eric Imsand
Editors

National Cyber Summit (NCS) Research Track 2021

Springer

Editors
Kim-Kwang Raymond Choo ⓘ
Department of Information Systems
and Cyber Security
The University of Texas at San Antonio
San Antonio, TX, USA

Tommy Morris ⓘ
Department of Electrical
and Computer Engineering
University of Alabama in Huntsville
Huntsville, AL, USA

Gilbert Peterson ⓘ
Department of Electrical
and Computer Engineering
Air Force Institute of Technology
Wright-Patterson Air Force Base, OH, USA

Eric Imsand ⓘ
Information Technology and Systems
Center (ITSC)
University of Alabama in Huntsville
Huntsville, AL, USA

ISSN 2367-3370   ISSN 2367-3389   (electronic)
Lecture Notes in Networks and Systems
ISBN 978-3-030-84613-8   ISBN 978-3-030-84614-5   (eBook)
https://doi.org/10.1007/978-3-030-84614-5

This Springer imprint is published by the registered company Springer Nature Switzerland AG
The registered company address is: Gewerbestrasse 11, 6330 Cham, Switzerland

Preface

While governments around the world have focused on strengthening their cybersecurity posture in recent years, cybersecurity remains a topic of ongoing importance. For example, in the "Executive Order on Improving the Nation's Cybersecurity (May 12, 2021)[1], it was reported that:

> *The United States faces persistent and increasingly sophisticated malicious cyber campaigns that threaten the public sector, the private sector, and ultimately the American people's security and privacy. The Federal Government must improve its efforts to identify, deter, protect against, detect, and respond to these actions and actors. The Federal Government must also carefully examine what occurred during any major cyber incident and apply lessons learned. But cybersecurity requires more than government action. Protecting our Nation from malicious cyber actors requires the Federal Government to partner with the private sector. The private sector must adapt to the continuously changing threat environment, ensure its products are built and operate securely, and partner with the Federal Government to foster a more secure cyberspace. In the end, the trust we place in our digital infrastructure should be proportional to how trustworthy and transparent that infrastructure is, and to the consequences we will incur if that trust is misplaced.*

As we have noted in the past years, there is a continuing need to keep a watchful brief on the cyber threat landscape, and this is the intention of this conference proceedings.

This conference proceedings contains a total of 13 papers consisting of both regular and invited papers from the 2021 National Cyber Summit Research Track. The 2021 National Cyber Summit was originally planned to be held in Huntsville, Alabama, from June 8 to 10, 2021. However, due to the COVID-19 pandemic, all tracks of the 2021 National Cyber Summit were delayed until September of 2021. The 2021 National Cyber Summit Research Track was held in-person from September 28 to 30. Authors from each selected paper presented their work and took questions from the audience.

---

[1]https://www.whitehouse.gov/briefing-room/presidential-actions/2021/05/12/executive-order-on-improving-the-nations-cybersecurity/.

The papers were selected from submissions from universities, national laboratories, and the private sector from across the USA. All of the papers went through an extensive review process by internationally recognized experts in cyber-security.

The Research Track at the 2021 National Cyber Summit has been made possible by the joint effort of a large number of individuals and organizations worldwide. There is a long list of people who volunteered their time and energy to put together the conference and deserved special thanks. First and foremost, we would like to offer our gratitude to the entire Organizing Committee for guiding the entire process of the conference. We are also deeply grateful to all the Program Committee members for their time and efforts in reading, commenting, debating, and finally selecting the papers. We also thank all the external reviewers for assisting the Program Committee in their particular areas of expertise as well as all the authors, participants, and session chairs for their valuable contributions.

Tommy Morris
Kim-Kwang Raymond Choo
Gilbert Peterson
Eric Imsand

Organization

Organizing Committee

General Chairs

Tommy Morris — The University of Alabama in Huntsville, USA
Kim-Kwang Raymond Choo — The University of Texas at San Antonio, USA

Program Committee Chairs

Gilbert L. Peterson — Air Force Institute of Technology, USA
Eric Imsand — The University of Alabama in Huntsville, USA

Program Committee and External Reviewers

Program Committee Members

Cong Pu — Marshall University, USA
Jun Dai — California State University, USA
Ezhil Kalaimannan — University of West Florid, USA
David Dampier — Marshall University, USA
Robin Verma — University of Texas at San Antonio, USA
Jianyi Zhang — Beijing Electronic Science and Technology Institute, China
Patrick Jungwirth — US Army Research Laboratory, USA
Junggab Son — Kennesaw State University, USA
Reza M. Parizi — Kennesaw State University, USA
Jaewoo Lee — University of Georgia, USA
Vahid Heydari — Rowan University, USA
Yifei Wang — Alipay, USA
Wei Zhang — University of Louisville, USA

David Coe   University of Alabama in Huntsville, USA
Junghee Lee   Korea University, South Korea
Huijun Wu   Arizona State University, USA
Ravi Rao   Fairleigh Dickinson University, USA
Rongxing Lu   University of New Brunswick, Canada

External Reviewers

Einaam Alim
Raphael Barata
Pinyao Guo
Hussam Al Hamadi
David Hayes
Erdal Kose
Yaoqing Liu
Zach Tackett
Chunxu Tang
Benjamin Turnbull
Xiaolu Zhang
Shaohua Wang

Contents

Cyber Security Education

An Integrated System for Connecting Cybersecurity Competency, Student Activities and Career Building

Li-Chiou Chen[(⊠)], Andreea Cotoranu, Praviin Mandhare, and Darren Hayes

Pace University, New York, NY 10038, USA
{lchen,acotoranu,pmamdhare,dhayes}@pace.edu

Abstract. For educators, preparing students who are able to solve cybersecurity problems requires not only a curriculum that provides students with interdisciplinary knowledge but also activities that develop their skills and competencies to solve problems that call for an interdisciplinary approach. To facilitate this process, we have developed a system called Cyberpassport, which integrates students' academic and career goals with cybersecurity co-curricular activities. This system is designed for students to search and register for cybersecurity activities, and track their own progress, for advisors to support their mentoring efforts, and for activity hosts such as faculty members or industry professionals to facilitate connection with interested students. Usability testing was conducted to test the application's functionality as well as user experience and interest. Preliminary results indicate that users found the system easy to use and beneficial for a career in cybersecurity. Furthermore, this is the first mobile events app developed that aligns with the skills and competencies defined by the National Initiative for Cybersecurity Education (NICE) cybersecurity workforce framework.

Keywords: Cybersecurity education · Competency · Workforce · Usability

1 Introduction

Cybersecurity has emerged as an academic discipline because of its importance for organizations in the digital era, in addition to the sophistication of knowledge and skills needed for cybersecurity professionals. The Association for Computing Machinery's (ACM's) Joint Task Force on Cybersecurity Education [1] defined cybersecurity as "a computing-based discipline involving technology, people, information, and processes to enable assured operations in the context of adversaries. It involves the creation, operation, analysis, and testing of secure computer systems. It is an interdisciplinary course of study, including aspects of law, policy, human factors, ethics, and risk management." For educators, preparing students to solve cybersecurity problems requires not only a curriculum that provides students with the knowledge and topics that are interdisciplinary in nature but also activities that develop their skills and abilities to solve these interdisciplinary problems. In addition, connecting with cybersecurity professionals and learning from them are important steps in developing a career in cybersecurity.

© The Author(s), under exclusive license to Springer Nature Switzerland AG 2022
K.-K. R. Choo et al. (Eds.): NCS 2021, LNNS 310, pp. 3–12, 2022.
https://doi.org/10.1007/978-3-030-84614-5_1

To facilitate this process, we have developed a system called Cyberpassport, which integrates students' academic and career goals with cybersecurity co-curricular activities. This system is designed to enable students to search and register for cybersecurity activities, track their own progress, support advisors in their mentoring efforts, and facilitate activity/conference hosts (e.g. faculty members or industry professionals) to connect with interested students. In addition, the system will allow students to generate a resume using information captured in the system, including activities that are labeled with skills and competencies defined in the National Initiative for Cybersecurity Education (NICE) cybersecurity workforce framework [2], which they can ultimately share with potential employers.

Cyberpassport enables students to connect with professional development training or co-curricular activities. The system aims to integrate students' career goals with cybersecurity training sessions, while allowing students to connect to these sessions. Using either the Cyberpassport mobile app or the Cyberpassport website, users of the system can host, search, register and record these training sessions, and detail the skills/competencies that they have learned. Students can then track their progress, review their skill set with faculty mentors or academic advisors, or share the information with potential employers, like a resume on-the-go. The cyber skills utilized in the system aligns with the knowledge/skills outlined in the NICE cybersecurity workforce framework, which is helpful for students when planning for their cyber career development.

Industry professionals can also leverage Cyberpassport to organize cyber activities that can be searched by students. The system allows users to create an event, such as a training session or a workshop. After being reviewed and approved by the administrator of its organization, the event will become available and searchable by all of the registered users. The system also has the potential to be used by employers interested in finding cybersecurity talents via event hosting, if students opt to share their information with employers.

We would like to engage the cybersecurity community in adopting Cyberpassport. The system is scalable and can accommodate a diverse range of cybersecurity activities offered throughout the community. The value of the system will ultimately depend on the number of events that will be entered into the system, and the number of students who will adopt the system. Once the system achieves a critical mass, it will help students to identify and record cybersecurity activities related to the knowledge and skills they aim to strengthen, while assisting advisors seeking to guide students towards meeting their academic and professional goals. This system is unique in the way that it connects students with cybersecurity activities, and can contribute to strengthening the students' knowledge and skills needed for a career in the field.

2   Literature Review

Fostering student competencies in cybersecurity education is critical. The consensus amongst educators is that cybersecurity students should not only obtain theoretical knowledge but should also be trained with the skills and abilities to perform cybersecurity related tasks. NIST's NICE framework [2] defines "competency" as a mechanism for organizations or educators to assess student's overall ability to accomplish a prescribed project, which can be described as a set of tasks, knowledge, and skills. For instance, an example of a "data analysis" competency could be "The collecting, synthesizing, or analyzing qualitative and quantitative data and information from a variety of sources to reach a decision, make a recommendation, and/or compile reports, briefings, executive summaries, and other correspondence" [2]. The Department of Labor's Employment and Training Administration (ETA) has also defined a Cybersecurity Competency Model [3], which compliments the NICE framework by adding competencies required by the average worker who uses the Internet or an organization's computer network. The National Security Agency/Department of Homeland Security's designation of National Center for Academic Excellence in Cybersecurity (NCAE-C) has also emphasized the importance of competency in cybersecurity education.

Competency has also been defined as a qualification for jobs and assessed by an assessment body to provide industry certifications. Assessment organizations, such as CompTIA, EC-Council, have utilized many techniques to assess the competency of cybersecurity professionals. A previous study [4] analyzed these techniques used for industry certifications and the perception of employers and the efficacy of competency assessment techniques. The study found that multiple choice examinations, used in many certification exams, are perceived as the least effective assessment method, while the qualification bodies use these most frequently. Oral examination, virtual lab examination, employment history and a review of qualifications are perceived as more effective but are used less frequently in the assessment by a qualification body. Another study [5] surveyed cybersecurity professionals and educators and assessed the preparedness of cybersecurity students in terms of competency. The study highlighted the importance of workplace competencies, as perceived by both professionals and educators.

There is a chasm between educators and employers in terms of training students to establish and assess competencies. We designed the Cyberpassport system to bridge the gap so that students can use the tool to explore potential cybersecurity job roles, plan and build competencies needed for these roles, and use the system to document their achievements.

3 The Cyberpassport System

3.1 System Design

As shown in Fig. 1, Cyberpassport is designed as a Web service catering to activity hosts, students, faculty advisors, and employers. CPS is designed specifically for students interested in a career in cybersecurity and allows them to search and sign up for cybersecurity co-curricular activities, such as workshops in a conference, research discussions on campus, or cybersecurity competition training.

Although the approach is conventional in the design and the security features, the innovation of the system is the integration of the design with cybersecurity education, in particular the NICE cybersecurity workforce framework. The hosts of these activities (activity hosts) can use the CPS as a mechanism to promote their cybersecurity activities by adding them to the system and labeling them with the knowledge/skill information defined by the NICE cybersecurity workforce framework.

Moreover, faculty advisors can mentor students based on the types of activities available, activities attended, and the knowledge/skills that the students need in order to achieve their career goals.

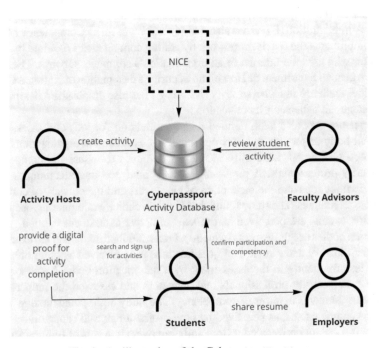

Fig. 1. An illustration of the Cyberpassport system.

Cyberpassport uses a method similar to challenge-response authentication, such as MS-CHAP or PPP-CHAP [6] but incorporates a mobile application to obtain some information required to validate student participation in activities. The system examines a *digital proof* from a student to validate participation. The activity host provides some validation when the activity is satisfactorily completed and the student provides yet another validation.

Activity hosts can register activity information, such as event date, time, duration, name, NICE competency identifier, and an associated URL that contains detailed information about the activity, such as an event website. Each activity is associated with an activity identifier which is stored in the system. A hash of the activity identifier, called an *activity hash*, is used later to confirm completion of the activity.

Students are able to browse and sign up for activities using a client-side mobile application. When the student creates an account in the system, the system generates a student identifier, which is encrypted and stored on the system. A hash of the student identifier, called the *student hash*, is sent to the student and stored on the end-user device, such as within a mobile application. When the student signs up for an activity, the system generates and stores a *digital proof*, which is a hash of the combined value of the activity hash, the student hash, and a password. The digital proof is only used by the system and is concealed from other users at this point.

When a student completes an activity and the activity host is satisfied with their participation, the activity host provides the student with the activity hash, implemented in the form of a QR code generated from the system. The students' client-side interface, either the Web interface or the mobile application, generates a digital proof using the activity hash obtained from the activity host, the student hash stored in their mobile application or Web interface previously, and the password that the student knows. This digital proof is sent to the system and the system validates the participation by comparing this digital proof with the one stored in the system previously. If these two hashes match, the student has a completion record, which shows the student has completed training on a specific NICE competency for a specific duration (e.g. two hours).

Students are also able to use Cyberpassport to generate a resume-like record that reflects the cybersecurity activities completed, and which can be used for employment purposes. Furthermore, advisors could use the system to review student progress and make recommendations for further academic or professional development.

3.2 Implementation

We have implemented Cyberpassport[1] both as a Web application as well as mobile applications for both iOS and Android. The Web application provides access to activity hosts, faculty advisors and students. The mobile applications are developed for students to browse the cybersecurity activities, scan the activity hashes after completing an activity, and send a digital proof to the system.

User Interface. Both a mobile app and a website (Fig. 2) have been developed. The front-end user interfaces, for Web and mobile, are connect to a MySQL database hosted on a Linux server at our university. Users can use either the website or the mobile app to access the system. Both interfaces provide the same functionality.

Cyberpassport QR Code. Once a user creates an account, the user is assigned a unique identifier, called a Cyberpassport number, which can be displayed as a QR code.

---

[1] Cyberpassport is available as a Web application at cyberpassport.pace.edu, and as a mobile apps at both Apple Store and Google Play.

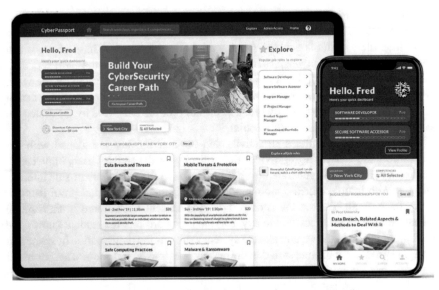

Fig. 2. Cyberpassport Web application and mobile application.

Cybersecurity Event Hosting. A user can create a cybersecurity event, which is defined as a cybersecurity skill training session, for one to three hours (Fig. 3(a)). Each event needs to be a physical or virtual gathering, with an actual date/time and location. Additionally, each event is required to list competencies or skills that are defined in NIST's NICE Framework. Once an event has been created, it will be reviewed by the Cyberpassport system administrator to ensure that the event is appropriately labeled and that its content is cybersecurity related. Once the event is approved by the administrator, users can then register for the event.

Event Registration. A user can search for cybersecurity events based on NICE skills or other keywords (Fig. 3(b)). Once a desired event is identified, the user can register for it and the host will have access to the list of event attendees.

Event Participation. The event host will scan the user's Cyberpassport QR code only if the participant attends the entire event.

Cyberpassport Skills. Users can review and demonstrate the skills that they have acquired from various events using their Cyberpassport mobile app. Users can also explore cybersecurity job roles defined in the NICE framework (Fig. 3(c)).

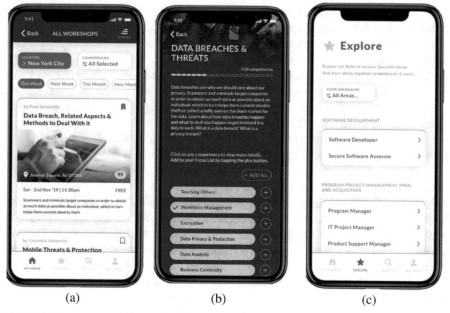

Fig. 3. Cyberpassport mobile applications. 3(a) Cybersecurity event; 3(b) Search events using NICE Competency. 3(c) Explore cybersecurity job roles.

4 Usability Testing

To determine whether Cyberpassport was easy to use, understandable, and satisfied the end user needs, we conducted four rounds of user studies. Of these user studies, three were conducted for the Web application, and one was performed for the mobile application. We reiterated the Cyberpassport interface design based on end user feedback, following each round of testing. In addition, we designed a questionnaire to collect data and to evaluate the end user's perception in using the application based on the Technology Acceptance Model [7]. The questionnaire included a combination of open ended and multiple-choice questions. The questionnaire focused on usability as it pertains to key application functions as well as evaluating the end user's perceived usability and perceived ease of use when using Cyberpassport.

We created and hosted a series of events to provide context to the user studies, and gathered user feedback on the application. These events included three workshops on the topic of "*Cybersecurity Analytics with Python,*" and one presentation on "*Cybersecurity Careers: Knowledge, Skills and Abilities.*". The events were promoted through the school and Cybersecurity Club social media channels, which connect with undergraduate and graduate students at our institution. The Cybersecurity Club students were a relevant end-user population because of their affirmed interest in cybersecurity. The total number of event participants was 68, and we collected a total of 53 complete responses. Of these responses, many were from participants in more than one study.

The event invitations directed students to create an account with Cyberpassport, and to register for each event using the application. All tests were conducted virtually, over

Zoom. The participants were prompted to explore the Cyberpassport application on their computer or mobile device, advance through a set of tasks, and answer a questionnaire. The test questionnaire was administered through Qualtrics A test proctor explained the scope of the study and provided instructions.

The participants were asked to reflect on the Cyberpassport application based on the usefulness, ease of use, and intent to use. In terms of "usefulness", participants generally agree that the application helped with searching for events. One participant commented *"I think it is useful to find what workshop I can register for."* In terms of "ease to use", participants generally agree that the functionality is accessible and organized but there are minor formatting issues that need to be fixed. One participant commented *"Cyberpassport is relatively easy to use. There are some errors in the formatting that cause the website to slightly break, but other than that it is easy to use."* In terms of "intent to use", although most participants expressed the intention to use the application in the future, some indicated their participation will depend on the activities offered on the application. Sample comments include *"Yes because it is probably going to be where I can access many of the resources I want regarding cybersecurity education"* and *"Don't know, possibly, I did like it but need to know more what is for."*

While we understand that many mobile applications share data with third-party analytics companies, we decided not to collect this data to respect the privacy of our users, while maintaining compliance with privacy legislation, such as the General Data Protection Regulation (GDPR) and the California Consumer Privacy Act (CCPA). Rather, we have relied on our own usability testing and surveys.

5 Discussion

Although our usability testing provides preliminary evidence that the Cyberpassport application is easy to use and is perceived as useful among the test participants, the application will require a critical mass of users to sustain a supporting community. To help students in building their cybersecurity career paths, the supporting community will include the activity hosts who would use the system in offering cybersecurity events that are mapped to the competency defined in the NICE framework, the students who would use the system to search for events that can improve their cybersecurity competencies, and the employers who would take advantage of the information provided by the system to identify talent.

We intend to integrate the Cyberpassport application with the Cybersecurity Club activities at our institution. This integration will support further dissemination of the application among the college student population and support its development further. In addition, the application will provide a place for students to advertise club activities such as student-led cybersecurity workshops, competition practice, and certification study group.

Currently, the Cyberpassport system does not provide a cybersecurity curriculum but rather serves as a system for career path exploration through events registered with the system. However, cybersecurity curriculum materials and tools [8–11], like the NICE Challenge, Labtainers, SEED and teaching modules on Clark, etc., can utilize Cyberpassport to create events and to advise students on skills or competencies required to pursue a specific cybersecurity career path. For example, the host of a penetration testing workshop can use the system to recruit participants, and the participants can find the workshop in the system if they are looking to build competency in penetration testing. However, the host will need to use another platform to host the workshop, either in a physical location or virtually. The Cyberpassport system can store links to materials used in a workshop, such as lessons and exercises, but cannot store these materials.

6 Conclusions

We have designed and implemented an integrated system to connect students with cybersecurity events or co-curricular activities. Using this system, students can explore cybersecurity job roles while searching for activities to help them in building competencies for careers in cybersecurity. This system is unique because it connects students with cybersecurity activities and can contribute to strengthening the students' knowledge and skills required for a career in the field. We believe that Cyberpassport can be used as a tool for mentoring students and preparing them for careers in cybersecurity with the support of the community.

Acknowledgement. The authors would like to acknowledge the support from the U.S. Department of Defense under Grant No. H98230–18-1–0309 and No. H98230–19-1–0293. Any opinions, findings, and conclusions or recommendations expressed in this material are those of the author(s) and do not necessarily reflect the views of the Department of Defense, or the U.S. government.

References

1. Cybersecurity Curriculum Guidelines. The Association for Computing Machinery (ACM). Joint Task Force on Cybersecurity Education (2017)
2. NICE framework. https://nvlpubs.nist.gov/nistpubs/SpecialPublications/NIST.SP.800-181r1.pdf
3. ETA, Cybersecurity Competency Model. https://www.careeronestop.org/competencymodel/competency-models/cybersecurity.aspx
4. Knowles, W., Such, J.M., Gouglidis, A., Misra, G., Rashid, A.: All that glitters is not gold: on the effectiveness of cyber security qualifications. Computer **50** (2017). https://doi.org/10.1109/MC.2017.4451226
5. St. Clair, N., Girard, J.: Judging competencies in recent cybersecurity graduates. J. Colloq. Inf. Syst. Secur. Edu. **8**(1), 10 (2020)
6. Microsoft PPP CHAP Extensions, IETF RFC 2759. https://tools.ietf.org/html/rfc2759
7. Davis, F.D.: Perceived usefulness, perceived ease of use, and user acceptance of information technology. MIS Q. **13**(3), 319–340 (1989)
8. Thompson, M.F., Irvine, C.E,: Individualizing cybersecurity lab exercises with labtainers. In: IEEE Security & Privacy, vol. 16, no. 2, pp. 91–95, March/April 2018. https://nps.edu/web/c3o/labtainers
9. Du, W.: The SEED project: providing hands-on lab exercises for computer security education. In: IEEE Security and Privacy Magazine, September/October, 2011. https://seedsecuritylabs.org/

10. Nestler, V., Coulson, T., Ashley, J.D.: The NICE challenge project: providing workforce experience before the workforce. In: IEEE Security & Privacy, vol. 17, no. 2, pp. 73–78, March-April 2019. https://doi.org/10.1109/MSEC.2018.2888784. https://nice-challenge.com/
11. Dark, M., Kaza, S., Taylor, B.: CLARK – the cybersecurity labs and resource knowledge-base – a living digital library. 2018 USENIX Workshop on Advances in Security Education (2018)

Simulating Industrial Control Systems Using Node-RED and Unreal Engine 4

Steven Day[1]($\boxtimes$), William "Kohler" Smallwood[2], and Joshua Kuhn[3]

[1] Argonne National Laboratory, Lemont, IL 08544, USA
days@anl.gov
[2] Mississippi State University, Starkville, MS 39759, USA
Wks68@msstate.edu
[3] Moraine Valley Community College, Palos Hills, IL 60465, USA
kuhnj8@student.morainevalley.edu

Abstract. The cost of building industrial control systems for cyber security research is a barrier far too high for most research institutions let alone space and safety requirements, leading the field to search for cheaper alternative solutions which may not always provide the same benefits. Through real world use cases, Argonne National Laboratory staff and interns developed a novel method of simulating industrial control system's environments using Node-RED, an open-sourced software, to handle programmable logic controllers' tasks and update and host a Modbus TCP/IP server. Through the CyberForce Competition™, the team has tested the usability of this method and has created 3D reactive models and physical models that run based the systems simulated in Node-RED. This method of simulation allows for easily deployed industrial control systems for any researcher to use in simulating how it might respond to attack, allowing for low budget institutions to do cutting edge research in the field of industrial control system security.

Keywords: Industrial control systems · Node-RED · Programmable logic controllers · Security · Simulation

1 Introduction

The price to build industrial control systems (ICS) for research is extremely high for most research laboratories. The goal of the work within this paper is to discuss a way to use open-source software to build an ICS simulation that would allow for ICS research without the cost of the overhead, or the risk involved with working with hardware. Using Node-RED, this can be achieved, allowing for lower budget institutions to perform higher yield ICS research. Also, by combing Node-RED with both scaled physical models and logical applications, creating a fully simulated ICS no longer requires a dedicated lab environment. Leveraging visualization software, like Unreal Engine 4, will add tremendous perception into to direct reactive state of these cyber-physical systems. This paper is broken down into the current problem, real world example, and proposed solution.

© The Author(s), under exclusive license to Springer Nature Switzerland AG 2022
K.-K. R. Choo et al. (Eds.): NCS 2021, LNNS 310, pp. 13–21, 2022.
https://doi.org/10.1007/978-3-030-84614-5_2

2 Industrial Control System Problems

Programmable logic controllers (PLCs) are devices that make linear decisions based off sensor input being fed into them from various inputs. These devices are expensive, and once they are put in place, tend to not move from their original purpose. For example, the SCADAPACK RTU PLC is approximately $2000 refurbished [2]. These devices also use proprietary software for each manufacturer's version of a PLC, leading to new knowledge needed to run any new kind of device, making it difficult to upgrade PLCs in existing systems and difficult to research general security practices on all PLCs of similar logic. The last two problems are space limitations and monetary limitations, PLCs take up physical space in a room, with physical models showing the decisions, and physical sensors feeding them data. This is what they are designed to do, and it normally does not pose a problem, however when wanting to research many PLCs that space can be filled up quickly. Finally, some security practices may provide excellent security to devices, but in testing might break certain portions of a PLCs connection, or the PLC itself. When researching these PLCs their price tends to limit the amount of testing researchers are willing to do when the potential to brick the device from either firmware mishaps or electrical miswirings, come into place, playing into the problems of monetary limitation.

3 Industrial Control System Prototypes Within Educational Modules - CyberForce Competition™

The CyberForce Competition™ is an annual Red/Blue team event held by the Department of Energy National Laboratories, originally started by Argonne National Laboratory. The competition focuses on defending ICS devices, as well as managing an enterprise style network infrastructure, all while defending from a barrage of attacks by a trained red team. Throughout the competition blue team members must think on their feet and develop new ways to prevent red team infiltration and maintain services their customer base might need even while under attack.

This competition holds interest for two reasons: 1) it allows for students to understand all portions of the confidentiality, integrity, and availability (CIA) triad, defending a system while still being forced to provide availability to their consumers, and 2) provides an insight into why ICS and operational technology (OT) security is so important in today's cyber landscape. However, providing an actual PLC for students to defend is extremely costly and complicated to handle especially with recent COVID-19 restrictions on in person activities. So, the problem arose, how might you simulate an ICS, while only being able to use basic information technology (IT) based programming languages? And when thinking about this issue, it became clear this problem was a concern outside of just the scope of the competition but could potentially benefit ICS research in different capacities.

4 Proposed Research Solution

Due to the rapid need of the CyberForce Competition™ and the needs seen through various ICS research at our institutions, particularly in PLCs, a method of simulating

ICS environments was prototyped. This new method utilizes virtual machines using VirtualBox Hypervisor on Windows 10, Node-RED, and Python and MySQL databases to build a fully functioning ICS environment, feeding sensor data through a JSON formatted set, allows the PLCs to make decisions like normal, and allows a user to use the human machine interface (HMI) to take control of the system as deemed necessary. This system is unique among similar systems due to its use of Modbus and Unreal Engine 4 to visualize the PLC decisions.

4.1 Node-RED Software

Node-RED is a tool created with NodeJS and JavaScript to provide a simple way to combine JavaScript logic and Internet of Things (IoT) devices [1]. With a large community building libraries for various functions and a simple interface to connect to hardware, it is a great choice for simulating Industrial Internet of Things (IIoT) and PLC type devices.

4.1.1 Data Flows

Node-RED works in programs called, "Flows". These flows carry messages through "Nodes" that do various types of computations on them. For example, if there was a lightbulb that was to be turned on using a JSON string "{bulb: on}", a flow would carry that message, a Node would read the message, and then the flow would continue to turn on or off, based on the message. This means, 1) We can send in sensor data and build JavaScript functions that will determine the state of a PLC and 2) JSON can operate like a normal IoT device, while still maintaining the same level of input an actual PLC would receive. Using this flow functionality, a flow can be built that receives sensor data, parses that sensor data, and then updates the PLC state to whatever that data would originally mean for an actual PLC, therefore simulating PLC functionality in an IT space.

These flows can also use protocols like Modbus and Transport Control Protocol (TCP) to send messages across machines. Utilizing a Modbus library created by the Node-RED community, Modbus TCP/Internet Protocol (IP) was used to simulate a PLC using the same protocol. All sensor data and decisions were stored on one device, while another flow, the command and control (C&C), took those decisions and showed them on an HMI window, which allows for operators to view current states and provide manual overrides. This means that a flow can be created for each portion of an ICS system: PLC, C&C, HMI, and a data historian, leaving a considerably basic, but equally as powerful ICS system that can be used to simulate attacks, show how defenses may work, and mitigate the risks of trying new things on expensive hardware like PLCs.

4.1.2 Modbus TCP/IP

Utilizing Modbus, an OT protocol used commonly in PLCs, was a large reason this system so closely resembles the outputs of an actual PLC decision making process. Modbus utilizes registers, which can be used to store values. These registers are given an address which allows for the values in them to be received and changed. Much like a linked list or other storage data structures, it allows for the user to store data they will need later and host it on an IP address to be easily accessible to other machines on the local network. The library "node-red-contrib-modbus" created by Klais Lansdorf, was used in this process. [4] This library generates a Serial Modbus server on the flow, which allows for the registers to be updated like a normal Modbus server would. This server can then be accessed by any machine on the local network using TCP connections. This ease of use allows for anyone to be able to build a Modbus server and communicate between two separate machines.

4.1.3   MySQL

To simulate a data historian server, all decisions made by the PLC needed to be updated to a database hosted on a separate machine. This database was communicated to a library called "node-red-node-mysql" which just reframes the NodeJS package to talk with a MySQL database [3]. Any changes to the Modbus server were sent as an update to the data historian to look at in the future for research and debugging purposes.

4.1.4   Dashboard UI/HMI

To build the HMI software that displayed the current sensor data, and allowed for override controls, the library "node-red-dashboard" was used [5]. Using this library allows for the building of controls into a separate hosted site, which can operate in place of an HMI. Running off HTML and JavaScript allows for this site to implement any JavaScript widget or input device and display information. This dashboard was the main output of the PLC system. It displayed all current sensor data and gave all override controls to the users needed for maintenance.

4.2   Modeling

4.2.1   Unreal Engine 4

A heavily missing aspect with many ICS simulation implementations is a real-time realistic and reactive visualization. Having the ability to display a direct response to a logical output, such as a malicious cyber-attack or an accidental engineer misconfiguration, can help to associate the effects of these system actions and allow for analysis and remediation training. Creating realistic renderings and animations from Unreal Engine 4 provides an aesthetic solution to these issues.

Unreal Engine, created by Epic Games, is a development suite most known for video game development, but also for animation, simulation, rendering, etc. [7]. Block-based module programming, called "blueprints", within Unreal Engine allow for a more human-readable approach to variable and socket creation [9]. With this, linking of the Node-RED software to communicate over network protocols to provide instantaneous sensor data delivery was easily obtained. Once the sensor data is received, real-time changes taking place on the PLC adjust variables in the simulation which, in turn, is reflected in the animation of the environment.

Due to the intensive resource need for running the Unreal Engine visualization, a machine with dedicated graphics, random access memory (RAM), and processing power was needed to achieve the full effect of the simulation. Leveraging a personal machine built for gaming, we were able to run the simulation at its full graphics potential and capacity (Fig. 1).

Fig. 1. Unreal Engine 4 visualization created for wind turbine example

4.2.2 Physical Model

Implementing a physical representation, to work in parallel with the Node-RED software with real-time results, was a desired addition to the need for the method prototype. Leveraging the Raspberry Pi 4 devices and their integrated general-purpose input/output (GPIO), we were able to replicate communications and actions of a PLC within the system [6]. A Raspberry Pi is an inexpensive and small single-board computer with variable RAM sizes and up to a gigabit speed networking interface. Taking advantage of a small and inexpensive device aids the limitations stated prior of industry PLC size and monetary value.

Utilizing Node-RED software to communicate between various ICS-based VMs and the Unreal Engine 4 simulation, sensor data was able to be used with system logic flows to make realistic actions and reactions. These logic-based actions and reactions cause electrical actuation to physical components, such as relay boards, power, motors, lights, etc. The wind turbines were modified from the original state of the LEGO™ Vestas wind turbine sets to connect and communicate to the PLC commands through the relay switch. By rewiring the motors, we were able to control the actuation from the relay boards, which in turn were controlled by the PLC. Additionally, to add more realism to the model, an aircraft warning light beacon system was appended onto the turbine. This was wired into the relay board and PLC as an additional component to be manipulated by the ICS.

The physical build was designed so that each of the four wind turbines had their own $2' \times 2'$ base giving a total landscape of $4' \times 4'$. Each turbine base had a dedicated

Raspberry Pi and 8-channel relay board attached. The Raspberry Pi was connected via Cat6 Ethernet cable and power supply, along with the GPIO cables to the relay board. The relay board was wired with an external power supply to deliver power to the turbine components, motors, and lights. All base Raspberry Pi units were connected to an unmanaged switch sending and receiving communications through the host machine of both the C&C VM and the data historian VM and to the Unreal Engine host machine (Fig. 2).

Fig. 2. Physical model of the wind turbine farm.

4.3 Wind Turbine Simulation

The first use case of this prototype system was revealed live during the 2020 CyberForce Competition™ as a small-scale wind turbine farm ICS that had a consistent uptime duration of over 48 h. The developed system consisted of VMs running the Node-RED software networked to four Raspberry Pi device PLCs, connected to multi-channel relay boards controlling various electrical components that make up the physical models of wind turbines.

VMs created and used in the ICS infrastructure included a C&C machine, a data historian machine, and a PLC machine. The C&C VM, running a Linux operating system (OS), was used as the HMI that would allow for users to interact with the control systems and the PLC by various switches and buttons integrated into the system logic. Changes made to the switches and buttons would have a direct effect on the rest of the system based on their integration logic. The HMI also contains a multitude of output graphs and graphics pertaining to the current state of the PLCs, turbines, and the sensor data, as shown in Fig. 3.

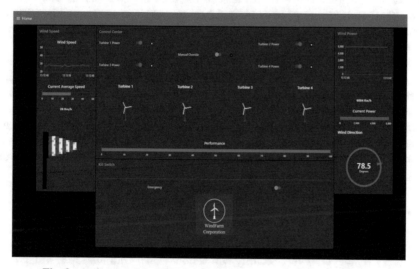

Fig. 3. An image of the HMI in use during the wind turbine simulation.

The data historian VM, also running a Linux OS, was developed with a Linux, Apache, MySQL, PHP (LAMP) stack back-end and phpMyAdmin as the web front-end to the database. The data historian stored all relevant data into various tables from all aspects of the system, including the C&C HMI changes, network connection activity between devices, sensor data feed from the PLCs, and turbine component state.

Finally, using the flows, the PLC VM was created using Node-RED flows and Modbus protocols. This PLC would take an input of JSON sensor data, updated regularly by a python script in the background, and feed it through the PLC to change the state of Modbus, and thus change the state of the machine. For the specific Modbus sensor data on the wind turbine simulation, data like wind speed, wind direction, and total power generation were sent via JSON, and used to determine the locked or unlocked state of the windmills.

This proof-of-concept simulation gave insight into how this system can be utilized cheaply to build a scalable version of ICS software. Throughout the implementation a few challenges where faced that needed to be overcome, the use of Unreal Engine 4 to model the system revealed graphics resource requirement problems that could not be met by a machine without designated graphics capabilities. Also, the environment needs to leverage the use of specific network address translation (NAT) port forwarding to communicate properly with the use of VirtualBox. Despite these drawbacks however, the system provided a fully visualized and functioning simulation of an ICS, which responds to attacks in real time.

5 Future Research

Some future research in the works for the system is completion of the attack-platform currently underway. This platform will allow for both automated and manual attacks to be performed against the ICS. As the attack platform is contributed to, a more diverse database of attacks is added to test against the differing subsystems. The use of an attack-platform is to attempt to quantify the cause-and-effect analysis derived from the reaction-based environment. Being able to quantify data can aid in the progression of ICS cyber security.

Alongside the attack platform, additional ICS critical infrastructure sectors are being designed to be modularly connected giving way to another research avenue of critical infrastructure dependency chains and the cascading effect cyber-attacks have on them. As seen in recent events of the 2021 Texas winter storm, critical infrastructure was taken down due to a natural storm. This natural storm rendered almost an entire state region incapacitated and without power. The adverse effects of this loss trickled down into many other facets of infrastructure, such as healthcare, transportation, water and wastewater, etc. [10].

6 Conclusion

Requiring a very swift shift from a typical in-person OT cyber security competition to completely remote OT cyber security competition came with a steep learning curve. This challenge not only affects students but it goes back to the researchers and operators of large-scale ICS equipment. Without the ability to be hands-on with equipment, large vulnerabilities continue to emerge. Utilizing this event to ignite this research avenue, a possible solution to a rising problem of how to create a complete ICS environment simulation that did not take up a large footprint and could be created at a lower monetary value came to light. Exhibiting a fully functional prototype to provide a more realistic approach to ICS simulation and modeling proves to be an important first step to further research and development in the field of cyber security. Being able to run the system continuously for a longer than intended duration and have a real-time data stream between all components, while producing the related cause and effect displayed that the prototype was a success.

Acknowledgement. Argonne National Laboratory's work was supported by the U.S. Department of Energy, Office of Science, under contract DE-AC02-06CH11357.

References

1. Node-RED Homepage. https://nodered.org. Accessed 6 Mar 2021
2. Control Microsystems SCADAPACK sale. https://www.radwell.com/en-US/Buy/CONTROL%20MICROSYSTEMS/CONTROL%20MICROSYSTEMS/P200-1A00-AA00. Accessed 6 Mar 2021
3. Node-red-node-MySQL. https://flows.nodered.org/node/node-red-node-mysql. Accessed 6 Mar 2021

4. Node-red-contrib-modbus. https://flows.nodered.org/node/node-red-contrib-modbus. Accessed 6 Mar 2021
5. Node-red-dashboard. https://flows.nodered.org/node/node-red-dashboard. Accessed 7 Mar 2021
6. What is a Raspberry Pi. https://www.raspberrypi.org/help/what-%20is-a-raspberry-pi/. Accessed 6 Mar 2021
7. Unreal Engine Features. https://www.unrealengine.com/en-US/features. Accessed 6 Feb 2021
8. Virtual Networking. https://www.virtualbox.org/manual/ch06.html#natforward. Accessed 6 Mar 2021
9. Blueprints Visual Scripting. https://docs.unrealengine.com/en-US/ProgrammingAndScripting.Blueprints/index/html. Accessed 6 Mar 2021
10. Power failure: how a winter storm pushed Texas into crisis. https://apnews.com/article/houston-football-storms-coronavirus-pandemic-hurricanes-5fd491ed5bfd9aa0ae08426c6078539e. Accessed 7 Mar 2021

Student Educational Learning Experience Through Cooperative Research

Melissa Hannis[1]([⊠]) [ID], Idongesit Mkpong-Ruffin[2], and Drew Hamilton[3]

[1] Center for Cyber Innovation, Starkville, MS, USA
melissa@cci.msstate.edu
[2] Florida Agricultural and Mechanical University, Tallahassee, FL, USA
[3] Mississippi State University, Starkville, MS, USA

Abstract. Two state universities are working together on a two-part research project. The first part revolves around creating a learning experience for these universities' students where they will develop key cybersecurity skills. They are tasked with creating and assessing fictional businesses to produce mock data that will be vital to the second part of this project. This will require them to develop a secure network architecture for their business along with security policies that consider the physical security of the organization not just the cybersecurity. The second part will be to design and develop a tool that will assist businesses in risk management. The created mock data will be used to test the tool during its development. We have hired students from both universities to work together to create several fictional businesses. These fictional businesses, once created, will be assessed using the Department of Homeland Security's Cyber Security Evaluation Tool (CSET). This tool produces a report from a business security assessment that will be used in the second part of this project. The student workers are split into groups, each of which have at least one student from each university with different levels of expertise to ensure that students have the chance for peer-to-peer learning and to network with students outside of their university. The goal for the first part of this project is to help students develop cybersecurity skills through cooperative research.

Keywords: Cybersecurity · Risk management · Business security · Collaborative learning environment

1 Introduction

Going to school for a computing degree and working within a computing practice are two very different learning experiences. Often graduating students have to "drink from a firehose" with the amount of knowledge that they must know going into a cybersecurity related job after graduation. The first part of this project revolves around creating data that will be utilized in the second part of the project, which is not discussed at length in this paper. The development of this data will provide a learning experience for the students where they will acquire key cybersecurity skills through cooperative research. These key skills include learning defense in depth, system hardening, how to develop a

K.-K. R. Choo et al. (Eds.): NCS 2021, LNNS 310, pp. 22–27, 2022.
https://doi.org/10.1007/978-3-030-84614-5_3

secure network, and how to develop security policies. This project is not only intended to create a possible end product but to provide students with cybersecurity and research skills they would not have been exposed to in their school courses.

Students are hired as paid interns and the work they are doing is utilized in the second part of the project. In the second part we are working to develop a Security Risk Assessment Tool (SRAT), and a Security Risk Assessment Model (SRAM), that will help companies better evaluate the risk to their company and provide guidance. This part of the project is still in the research phase and development has not yet begun. The proposed tool will give a quantifiable probability of a security breach for a company based on the NIST 800-171 controls and Cybersecurity Maturity Model (CMM) practices that are in place and how they have been implemented. This tool will be developed to automatically generate the probability of a breach, provided a Cyber Security Evaluation Tool (CSET) report for a given company. Each NIST 800-171 control and/or CMM practice that is implemented improves the security posture of the company and the probability of that company being breached is reduced. We will use natural language processing (NLP) and Machine Learning (ML) to train NIST 800-171 controls and CMM practices against the National Vulnerability Database (NVD) data, classify the controls and practices, and predict the score to be assigned to each company's posture based on the used classification models.

This tool will need a dataset as input to test and develop the tool. We are utilizing CSET, developed by the Department of Homeland Security (DHS), which allows companies to systematically assess the security posture of their company. CSET has an option to evaluate a company based on the NIST 800-171 controls or the CMM practices. These controls/practices are evaluated through a list of yes or no questions to determine if each is met. CSET encourages users to also include comments to explain how a control is met or how a control will be implemented. An important feature of CSET is that it compiles all the information given during the assessment into a report. However, currently due to a bug in CSET, the students do not have the option to create a CMM report with their comments. Therefore, they have been asked to assess their company on just the NIST 800-171 controls. CMM is based off the NIST 800-171 controls so this will not affect the development of our tool. There are several different reports to choose from, the CSET report that we want contains the comments made during the assessment. This includes if a control was implemented or not and how it was implemented which will be used by our tool to help determine the company's security posture.

It is not expected for a company to give out their CSET report for us to develop our tool, given there is inevitably confidential information in a CSET report. However, it is important for the development of this tool to have a dataset to test and develop its features. Therefore, we have hired students from both universities to work together to create several fictional businesses to produces the required input data. Their work involves creating, securing, and performing an assessment of their fictional businesses. They will assess their fictional companies with CSET, and those produced reports will be used as the needed input for our proposed tool. In this paper we focus on the first part of this project in which we go over the relevant research in this area, what is expected to be learned through creating the fictional businesses, and the research the students have put forth on this project.

2 Relevant Research

It is widely accepted that collaborative learning is an effective way of enhancing the assimilation of knowledge by students. In Micari & Pazos's study, it was shown that students who were involved in supplemental small-group peer-learning programs, had greater gains in "self-efficacy for course tasks and self-regulation for learning and they were less likely to use surface-level memorization in studying for the course" [2]. Researchers have noted that a student's learning success is highly affected by their ability to be actively engaged with the learning materials [3, 4] and effective use of strategy, which social interaction engenders and strengthens [5]. This use of active engagement has been shown to be more effective in garnering knowledge than traditional approaches such as lectures and group discussion [6]. Collaborative learning provides opportunities for students to work together to build a shared understanding and develop a stronger "inter-subjective meaning" in the subject matter under question [7].

Although most research in collaborative learning deals within the content of core curriculum [8] [6] within the classroom or with peers within the same cohort outside the classroom. This collaborative experience is uniquely different, in that it's diversity stems in partnering students from different schools, at different levels of matriculation through an IT, CS program in cybersecurity. They are able to use the synergies of their different perspectives to bring to bear on the learning experience.

3 Methodology

There are several different facets of any business and creating a realistic fictional business just on paper is still a lot of work. We have divided up the work into five sections: Organizational Background, Organizational Physical Layout, Organizational IT Infrastructure, and Organizational Procedures and Policies. This breaks down the major sections of a business that we will need to address during an assessment of the NIST controls or the CMM practices. With our first group of students we have divided up the students into five groups, each group has at least one graduate student from one of the universities and one undergraduate student from the other university. The students are paired in this way to encourage peer-to-peer learning and provide a means to broaden their network.

We have selected a few different types of business that range from a retail store to a research facility. The types of business are not particularly important, but we wanted to make sure we had a verity that will ideally stress the limits of our tool. Each team is assigned a business type at random to develop a background of an already established company.

Once the student groups have finished creating the fictional businesses and have run a CSET assessment on their business, the students will be assigned a new group and a different business type to develop. Changing the dynamics of the groups will give the students a chance to network with others on the project, and to learn how to secure a different type of business. The types of businesses we have addressed so far are: Retail, Pharmaceutical, Manufacture, Research and Development, and Private Practice. To make these fictional businesses representative of real businesses we ask each group to research two or three companies that most closely represents the business they are creating and to model their business after those companies. This ideally will help create a more realistic business and therefore a more accurate dataset.

The Organizational Background section encourages the students to be creative and to have fun in making whatever business they want within the type of business they have been assigned. In this section the student will be developing skills in research and learning about their given business type. We have provided them with a guideline to ensure they hit on some general points that should be addressed within each business. This includes defining the owners of the business and deciding when their business was founded. The idea is for them to bring their business to life with a brief back story. They are also asked how big they plan to make their business by providing the number of people within the organization and where their business is located. A business is not clearly defined without knowing the goals of the business or the organizations mission statement. Lastly the students must decide what their business's product or service is and their intended audience. All that is listed here is important for the realism of the company and for the students to see a small glimpse of what goes into creating a business.

For our project it is also important for them to know their business's financial standing. When cybersecurity professionals are working to secure a business, the cybersecurity is the main focus, but the finical security of a business should not be forgotten. Ensuring a company is secure is important but the cost should never outweigh the threat. By knowing their business financial standing, they will have to consider what they can afford in way of cybersecurity for their organization without damaging their finical security. This is one of the lessons we are trying to impart in this section of the project.

In the Organizational Physical Layout section, we ask the students to create a physical floor plan for each location that is part of their business. For example, it is typical for a manufacturing site to have a corporate office separated from the manufacturing site. In this case there would be two required floor plans, one for the corporate office and one for the manufacturing site. Knowing the physical layout of their organization is important when considering how to implement some of the physical NIST controls and/or CMM practices.

The students were asked to model their companies after at least two or three real businesses and to research these companies to bring more realism to their own fictional company. Part of their research into similar companies includes researching the most realistic physical layout for their business. Physical security must be taken into consideration given this is a big part of the battle when it comes to securing a business. In this physical layout the students must consider all entrances in and out of the building, the layout of the different departments, the public and private areas of the organization, determine if the Information Technology Services (ITS) is outsourced or in-house, and consider how they have physically protected their organizations information systems.

In the Organizational IT Infrastructure section, we have asked the students to research and create a report of the IT equipment and software they will use in their organization. They are also asked to design a network architecture for their organization. This should include each location that is part of their organization and show how these locations are connected. For example, it would be ideal to encrypted network traffic between the cooperate office and its manufacturing site and this is something that should be noted in their organization's network diagram. In this section they are expected to learn how to research for IT equipment and learn how to design and secure a network architecture. We hope through this project that the students will walk away with a better understanding of what it takes to secure a business network through a verity of techniques.

In the Organizational Processes and Policies, we asked the students to review the NIST 162 Handbook creating policies procedures that are needed for their specific organization. Before the students can perform an assessment of their organization with CSET they must first review the NIST 162 Handbook, which address all the NIST 800-171 controls and provides a means to better understand the controls. Reviewing the NIST 162 Handbook will give them the chance to reflect on the work they have done in the previous sections and document the policies and procedures they have already considered but have not put into words. It is not unusual for a business performing an assessment of their organization to find that they are already compliant with many of the controls but have not yet documented the procedure or created a policy. This section is intended to make the process of assessing their business with CSET easier, by ensuring they have the needed material prepared and ready for the assessment.

Finally, the students will perform a CSET assessment of their organization. CSET first prepares the assessment by getting information needed for the end report. This includes naming and dating the assessment, getting the name of the business or the facility that is being assessed, and getting the location of the business. There are several assessment options, but for our project we will be assessing based on the NIST 800-171 controls and we will include the assessment of the network diagram for the organization. CSET will prompt for the type of business they are assessing and the value of the assets they are trying to protect in the assessment setup. Then they will be prompted to create or upload the network diagram of their business. Finely, the students can go through their CSET assessment of the NIST 800-171 controls. Once the students have finished the assessment, CSET will compile all the entered data into a report. This report along with the material that they have created for their business will be used to develop the SRAT tool and the SRAM model.

4   Results

After each student has turned in their business report, their policy documentation, and their CSET report, the students were given two post surveys. One of the surveys was used to gauge what they have gained from this experience and the another survey, which was anonymous, was used to determine what they liked about this cooperative learning experience and what we could do to improve this experience. In the first survey most felt that they have learned more about cybersecurity, how to write business policies, and how to design a network through this experience. Some of the students stated in the survey that they also developed skills in time management, team collaboration, and how to present research. The general consensus of the anonymous survey was that many students found this experience beneficial, fun, and all said they would recommend this learning experience to others. There were some comments on how time-consuming learning and assessing their company on the NIST controls was, which for many companies is a difficult part of running an assessment. Knowing these controls will make these students very valuable to any organization.

Outside of these surveys several students came back and said that they have had an immediate benefit from working on this project. Many of the students used what they learned in their job interviews. They said the interviewers were impressed with their knowledge of the NIST controls and the cybersecurity skills they have gained on this project. Furthermore, they have been hired for summer internships.

5 Summary

This effort is an experiential learning experience for the student workers. In the beginning of the project we interviewed each student and got a baseline of each student's knowledge level. These students were paired best to ensure that each group had a person with expertise in both computer science and business. Some of the students had expertise in both subject matter and others were just at the beginning of their college careers with little to no experience in either field. This project is intended to give all students the opportunity for peer-to-peer learning and to learn how to preform research and report what they have found. This project will not only give the students a chance to aid in a final product but has also proven to have an immediate practical benefit to the student.

6 Future Work

The version of CSET that was used during this project was 10.1.0. This CSET version is not able to provide a report with all the information that was entered during the assessment of the CCM practices due to a bug. In the latest version of CSET that has just recently been released, 10.2.0, this bug has been fixed. In future iterations of the students creating and assessing their business, they will be able assess their business based on the CMM practices. CMM is becoming the new standard, this is the reason we would like the students to evaluate their business on the CMM practice instead of the NIST 800-171 controls in the future.

References

1. Stoneburner, G., Goguen, A., Feringa, A.: NIST SP-300 Risk Management Guide for Information Technology Systems (2002)
2. Micari, M., Pazos, P.: Beyond grades: improving college students' social-cognitive outcomes in STEM through a collaborative learning environment. J. Learn. Environ. Res. (2020)
3. Chi, M.: Active-constructive-interactive: a conceptual framework for differentiating learning activities. Top. Cogn. Sci. 1, 73–105 (2009)
4. O'Donnell, A.M., King, A.: Cognitive Perspectives on Peer Learning. LEA, Mahwah (1999)
5. Doise, W., Mugny, G., Perret-Clermont, A.: Social interaction and the development of cognitive operations. Eur. J. Soc. Psychol. 5(3), 367–383 (1975)
6. Barkley, E.F., et al.: Collaborative Learning Techniques: A Handbook for College Faculty. Wiley, Hoboken (2014). ProQuest Ebook Central: http://ebookcentral.proquest.com/lib/famu/detail.action?docID=1745058
7. Heisawn, J., Cindy, K.J., Hmelo-Silver, E.: Ten years of computer-supported collaborative learning: a meta-analysis of CSCL in STEM education during 2005–2014. Educ. Res. Rev. 28 (2019)
8. Banks, D.: Collaborative learning as a vehicle for learning about collaboration. In: InSITE (Informing Science), pp. 895–903, June 2003

Digital Forensics Education: Challenges and Future Opportunities

Megan Stigall and Kim-Kwang Raymond Choo(✉) [ID]

Department of Information Systems and Cyber Security, University of Texas at San Antonio, San Antonio, TX 78249, USA
mestigall@protonmail.com, raymond.choo@fulbrightmail.org

Abstract. Digital technologies are becoming more engrained in our daily life and society (e.g., smart city and smart nation), and these technologies can be both the target of and tool used to facilitate a broad range of malicious cyber activities. This reinforces the importance of disciplines such as digital forensics. Digital forensics is a relatively new, multidisciplinary field with roots in traditional forensic sciences, as well as combining elements of technology, legal, social science, political science, criminal justice, and various other disciplines. Compared to the more established fields of computer science, information security, and cybersecurity, digital forensics is somewhat understudied. Though interest in the field is growing and the industry is in need of trained professionals, there are a multitude of challenges to overcome as it relates to education. There is a glaring lack of standardization and structure both in existing educational programs and for those developing new digital forensics programs. Existing programs are scattered among various colleges and departments, and there are significant gaps in the materials covered and relevant emerging technologies. Despite these challenges, the digital forensics field and its variety of subfields such as Internet of Things, mobile, cloud, network, and vehicle forensics have been steadily gaining academic interest and attention. Finding solutions and developing robust higher educational programs is a necessary step to improve the quality of digital forensics education and produce highly trained professionals with the skills required to detect, investigate and prosecute malicious cyber activities in civil litigations (e.g., corporate espionage), criminal investigations and national security investigations.

Keywords: Digital forensics · Digital forensic education · Computer forensics · Education · Curricula

1 Introduction

An increasing quantity of new developments and innovations in technology in recent years has caused digital information to be built into almost every aspect of daily life. With these advancements, cybercrime is quick to follow, which in part fuels an increasing interest in disciplines such as digital forensics. We also remark that the terms digital forensics, cyber forensics, and computer forensics are often used interchangeably in the literature although computer forensics is generally used in earlier days of the discipline.

© The Author(s), under exclusive license to Springer Nature Switzerland AG 2022
K.-K. R. Choo et al. (Eds.): NCS 2021, LNNS 310, pp. 28–46, 2022.
https://doi.org/10.1007/978-3-030-84614-5_4

This relatively new field of digital forensics is a diverse and often under-reviewed branch of technological and/or forensic science studies, particularly in comparison to more established fields such as computer science and information/cybersecurity. There are varying interpretations of what digital forensics actually is, but there is a general consensus that digital forensics is an area of investigations that combines elements of social, legal, and ethical considerations with technology. According to the US Computer Emergency Readiness Team (US-CERT), the definition of digital forensics is:

"the discipline that combines elements of law and computer science to collect and analyze data from computer systems, networks, wireless communications, and storage devices in a way that is admissible as evidence in a court of law" [25].

Increasing numbers of both hardware and software in various technological innovations also imply an increased number of targets for cyberattacks [3]. For example, according to the US Director of National Intelligence, cybercrime was the top national security threat even back in 2014 [9]. Digital forensics plays a vital part both in investigation and often prosecution of cybercriminals, but the field is struggling to keep up and expand with growth of technology and the needs of the industry. Because of this, robust digital forensic education is of the utmost importance to produce trained professionals with the skills and expertise to detect, investigate, and prosecute individuals and/or other threat actors involved in malicious cyber activities, such as corporate espionage, theft of intellectual property by insiders and external attackers, and advanced persistent threat (APT) actors. Therefore, in this paper, we will focus on digital forensics education. Perhaps due in-part to the newness of the field, there is a lack of standardization or credentialing available to universities and institutions that aim to teach the study of digital forensics. Subfields of digital forensics such as Internet of Things (IoT) forensics, mobile forensics, cloud forensics, and more are on the rise (e.g., autonomous or unmanned vehicle forensics). Within these subfields, there are gaps in the topics covered by existing digital forensics courses and emerging trends. Programs and courses vary widely, and education and education development is made even more challenging by the multidisciplinary nature of the field.

For this literature review, numerous recent articles regarding digital forensics education were examined. The keywords or criteria used to search for relevant articles include the following: "computer forensics education", "digital forensics education", "computer forensics curriculum", and "digital forensics curriculum". The search was first conducted using Google Scholar as an overview, before also searching the following libraries: IEEE Xplore/IET Electronic Library Online, ACM (Association for Computing Machinery) Digital Library, ScienceDirect, SpringerLink, and SSRN (formerly known as the Social Science Research Network). A number of related journals such as the Journal of Cybersecurity Education, Research and Practice and the Journal of Digital Forensics, Security and Law were also examined for relevant sources. At the time of writing, twenty-eight (28) sources have been identified and a complete bibliography can be found following this review. Of course, for such a recently developing field of study, time periods of publication are meaningful. Articles published in the year 2020 - from January through November - were explored first, before moving on to 2019 and so on, through 2015 and later. The majority of the articles examined were published in 2019, and all but two were published in 2015 and later. The two works from prior to 2015 were acquired to highlight the early days of digital forensics education: one is a preliminary study on computer forensics programs in higher education, and the other is the original report on developing an innovative digital forensic program that precedes relevant later papers from the author on the same subject. Table 1 summarizes some of these existing works.

Table 1. Examples of existing studies on digital forensics education: a snapshot

Year of publication	Curriculum development or analysis	Course development or analysis	General overviews	Specialized implementations	Total
2020	N/A	Hasan et al.	N/A	George	2
2019	Bashir and Campbell; Harneker and Stander; Karabacak et al.; Naqvi et al.; Roy et al.	N/A	Dafoulas and Neilson; Verma and Bansal; Zahadat	Belshaw; Delija et al.; Deshpande and Ahmed; Doherty et al.; Leung and Blauw; Seda et al.; Tang; Wu et al.	16
2018	N/A	Antunes and Rabado	Luciano et al.	N/A	2
2017	Drange et al.	N/A	Kiper	N/A	2
2016	Liu; Liu	N/A	N/A	N/A	2
2015	Bicak et al.; Palmer et al.	N/A	N/A	N/A	2
...					
2006	Liu	N/A	N/A	N/A	1
2005	N/A	N/A	Gottschalk et al.	N/A	1
Total	11	2	6	9	28

In the next section, we will briefly introduce digital forensics.

2 Digital Forensics

Literature concerning digital forensics education covers a variety of subtopics, including curriculum development, course development, and even specialized implementations of digital forensics in various educational settings. A plethora of options on how to design and structure digital forensics education are presented, and of course these bring even more opinions on the best way to do so. The one thing the entire industry seems to agree on is this: digital forensics is a fast growing and fast changing field, and standardization is essential for success. Success here pertains to both in an educational setting as well as in an eventual professional setting, as one of the main goals of higher education in the field is to secure job placement.

Digital forensics was once said to be a niche, but now it is more of an interdisciplinary subfield of already existing programs such as computer science, information technology, information security, cybersecurity, and other, more solidly established technology programs [20]. It is also important to understand that digital forensics is not necessarily completely centered on technology either, as it has its roots in traditional forensic sciences. In fact, some would argue that forensic medicine - that is DNA, fingerprints, or material analyses - is not dissimilar [19].

Academic and research interest in digital forensics is still an emerging occurrence even today. In fact, the very first computer forensics higher education program was not introduced until 2004 [26]. Even now, over sixteen years later, the debate centered on the best manner of including the study of digital forensics into higher education is ongoing. To begin to form an idea of structuring a digital forensics education program, it is first necessary to understand the scope of the field.

2.1 The Multidisciplinary Nature of Digital Forensics

Perhaps one additional fact agreed upon in almost all the relevant literature is as follows: digital forensics is a multidisciplinary study. In addition to technology, there are social, ethical, and legal considerations to recognize. Then there are the subfields of digital forensics such as Internet of Things (IoT) forensics, mobile forensics, network forensics, and more. And of course, the number of subfields will likely grow as technology progresses. Current digital forensics curriculums vary widely in terms of the types of courses required, the number of courses, and the overall scope of the curriculum. Some digital forensics programs have a heavy emphasis on the technological aspect, while others prefer to focus on criminal justice and law [30]. Overall, courses tend to be split into the following categories: computer science, mathematics, and statistics; information technology and cybersecurity; criminal justice, law, and policy; business and management; and ethics. Digital forensics courses and programs also fall under various colleges and departments depending on the university. For example, some universities offer digital forensics courses under their Colleges of Computer Science, Computing, or Informatics, while others house them under departments of Engineering, Criminal Justice, or even School of Informatics, Humanities and Social Sciences. With such a variance of curriculums and wide variety of course subjects, it is no surprise that there is a glaring lack of standardization within digital forensic education.

Some of these variances are captured below in Table 2, which consists of twenty-five (25) universities in the United States (US) that were categorized as being one of the top STEM schools or top digital forensics schools in the US, and also offer at least one course or program in digital forensics. Comparisons were made among university information gathered from the National Center for Education Statistics, Bachelor's Degree Center, Crimson Education, Forbes, Niche, Universities.com, and Carnegie Classification of Institutions of Higher Education [10]. Forty-two universities were selected from these lists, but only twenty-five schools offered digital forensics courses. Of the twenty-five schools shown below, all but eight are classified as R1 or R2 Institutions by Carnegie.

Detailed information from each school was taken from each individual university's course catalog, which can be found in Appendix A.

Among digital forensic education researchers, there is a general agreement that learning outcomes are arguably the most important factor when designing curriculums [7]. Since one of the primary goals of higher educational institutions is future job placement of students [26], learning outcomes must first be tailored to fit industry needs. Research and investigation into the current needs of relevant industry institutions is a valid starting point when preparing a curriculum. Workshops and other events that bring together educators, students, and professionals in the field have been successful in broadening communication on the topic [3]. Another important point is to examine what level of education is necessary for the jobs and positions in need of educated applicants. For instance, forensic technicians, practitioners, policy makers, researchers, and educators may all be heavily involved in the world of digital forensics, but they likely require different levels of formal education to be successful in those roles.

Several common challenges faced in existing forensics programs have already been observed. Frequent obstacles included lack of established curriculum standards, lack of digital forensics textbooks available, trouble finding qualified faculty and educators, computer lab design, and disagreement on the topic of prerequisite requirements. These challenges will be revisited in the Research Methodology section. The lack of standards, textbooks, and field experts is entirely unsurprising in such a young and ever-changing field [30]. Advancements and progress in digital forensics education is imperative and it is possible that those particular issues will begin to fix themselves as the field progresses. Also unsurprisingly, another one of the main complaints is that education efforts in general are inconsistent and ineffective. Educators still fail to agree on how to scope the study of digital forensics [23]. According to Gottschalk et al. [19], bachelor programs are "conspicuously dissimilar". And these challenges are broad in themselves, as the discussion is still focused on overall curriculum. Many of the same issues persist when the courses within these curriculums are examined. In some programs, digital forensics is still lumped in as a module in a cybersecurity or information security course, rather than being a standalone topic. Then, within already developed digital forensics courses, there are still discrepancies regarding materials covered and not covered. Perhaps the multidisciplinary nature of the subject matter makes digital forensics less of a niche subtopic of cybersecurity, but more of a broad, extensive field.

Though challenges are numerous and differing opinions are widespread, proposed solutions and ideas for change are both confident and hopeful. The desire for a standardized educational strategy and curriculum is almost universally shared, and educators recognize the need for additional faculty training and development of new, relevant digital forensics textbooks. As computer labs have been identified as a vital part of a digital forensics curriculum [12], the need for appropriate infrastructure and funding is also an important consideration to make when developing a program. Some have suggested reforms in admission criteria as well, mainly due to the sensitive nature of digital forensics. In fact, some programs will exclusively admit law enforcement personnel [19]. As Liu [26] mentions, rapid technological change will also influence future digital forensics programs: as new advancements are made, new courses will need to be developed to

Table 2. US universities that offer at least one course or program in digital forensics (information accurate as of 12 November 2020)

University (see Appendix A)	College/Department	Offered as a degree program?	Offered as a concentration?	Offered as a course?	Research classification
Massachusetts Institute of Technology	Professional Education	No, offered as a 5-day virtual course	No	3 out of 41 topics include forensics	R1
Rice University	Tech Boot Camps	No, offered as a 24-week boot camp	No	1 out of 20 topics includes forensics	R1
Michigan Technological University	College of Computing	No	Yes	1 undergraduate course and 1 graduate course	R2
Stevens Institute of Technology	Computer Science	No	No	1 undergraduate course	R2
Missouri University of Science and Technology	Computer Science	No	No	1 undergraduate course	R2
Georgia Institute of Technology	Professional Education	No	No	1 professional course	R1
Illinois Institute of Technology	Information Technology Management	Yes, Master of Cyber Forensics and Security	No	2 undergraduate courses and 2 graduate courses	R2
Carnegie Mellon University	College of Engineering	No	Yes, Cyber Forensics and Incident Response track	4 undergraduate courses	R1

(continued)

Table 2. (continued)

University (see Appendix A)	College/Department	Offered as a degree program?	Offered as a concentration?	Offered as a course?	Research classification
Case Western Reserve University	Boot Camps	No, offered as a 24-week boot camp	No	1 out of 20 topics includes forensics	R1
Purdue University	Polytechnic Institute	No	Yes, Cyber Forensics Specialization	1 undergraduate course and 4 graduate level courses	R1
Johns Hopkins University	Engineering	No	No	2 graduate level courses	R1
University of Michigan	College of Engineering and Computer Science	Yes, Bachelor of Science in Digital Forensics	Yes, Digital Forensics concentration	6 undergraduate classes	R1
Tuskegee University	Computer Science	No	No	1 undergraduate or graduate level course	Not classified
University of Tulsa	College of Engineering and Natural Sciences	No	No	1 graduate level course	R2
Drexel University	College of Computing and Informatics	No	No	1 undergraduate level course	R1
Florida State University	College of Criminology and Criminal Justice; Computer Science Department	Yes, Bachelor's in Cyber Criminology; Master of Science in Cyber Criminology	No	1 undergraduate or graduate level course	R1

(continued)

Table 2. (continued)

University (see Appendix A)	College/Department	Offered as a degree program?	Offered as a concentration?	Offered as a course?	Research classification
Keiser University	Criminal Justice Department	Yes, Cyber Forensics/Information Security, B.S	No	4 undergraduate level classes	Not classified
Christian Brothers University	School of Sciences	Yes, Cyber Security and Digital Forensics, B.S	No	1 undergraduate level course	Not classified
Champlain College	Online Degrees & Certificates	Yes, Bachelor of Science in Computer Forensics & Digital Investigations; Master of Science in Digital Forensic Science	No	8 undergraduate courses and 10 graduate level courses	Not classified
Robert Morris University	School of Informatics, Humanities and Social Sciences	No	Yes, Digital Forensics concentration option for Cybersecurity B.S	3 undergraduate courses	Not classified
George Mason University	School of Engineering, Department of Electrical and Computer Engineering	Yes, Digital Forensics, M.S	No	19 graduate level courses	R1

(continued)

Table 2. (*continued*)

University (see Appendix A)	College/Department	Offered as a degree program?	Offered as a concentration?	Offered as a course?	Research classification
Stevenson University	N/A	Yes, B.S. Cybersecurity and Digital Forensics; M.S. Cybersecurity and Digital Forensics; Certificate in Digital Forensics	No	5 undergraduate courses and 5 graduate courses	Not classified
Strayer University	Criminal Justice; Information Systems	No	Yes, undergraduate Computer Forensics concentration; graduate Computer Forensic Management concentration; Computer Forensic Management minor	2 undergraduate courses, 1 graduate course	Not classified
Utica College	N/A	No	Yes, undergraduate Network Forensics and Intrusion Investigation specialization; graduate Computer Forensics specialization	1 undergraduate course and 4 graduate courses	Not classified
University of Arizona	College of Applied Science and Technology	No	Yes, Defense and Forensics Emphasis	2 undergraduate courses	R1

keep pace with industry needs as cybercrime inevitably follows the opportunities new technology brings.

2.2 Topics in Digital Forensics Education

Examining current topics that are covered - in addition to important topics that are not covered - in existing digital forensics courses and curricula is the first step to understanding digital forensics education and the potential areas where improvement is necessary. As stated above, academic and research interest into the field of digital forensics is still relatively new, and standardizations have not yet been established. In fact, the first article that discussed the emerging need for digital forensics education was published in 2003 [26]. Liu notes that educators have responded well and begun working with tremendous effort into developing new courses and programs. Now, programs are beginning to be established, but ones including digital forensics as a topic vary widely as well, and often digital forensics is just one or two general-overview-type class(es) in a larger degree such as Computer Science, Information Technology, or related field. As seen in Table 2 above, even the academic units that house digital forensics inclusive programs are diverse; several different colleges or departments, from Sciences to Engineering to Informatics etc., have been designated to hold these programs or courses.

Before looking at the specific course data, first, general data was gathered about the types of programs offered. The charts below highlight the differences in programs from top STEM universities versus those from top digital forensics universities. In fact, up to 46.7% of some of these top STEM universities in the US do not offer any programs or courses in digital forensics at all. Figure 1 below lays out some of these differences in the levels of programs and courses offered by the top STEM schools, while Fig. 2 shows the breakdown in program types of those universities who do offer material on digital forensics.

As shown in both Figs. 1 and 2, it is apparent that higher educational degrees in digital forensics are still fairly rarely offered overall, but that a significant number of institutions at least offer courses in digital forensics. 60% of universities that have been noted for their digital forensics educational paths offer digital forensics degrees, while only 16.7% of STEM-recognized universities offer digital forensics degrees. Comparing these figures with Table 2 previously, it becomes apparent that even universities that offer degrees may only have a small number of digital forensic specific courses that are included in the degree plan. Champlain College, George Mason University, and Stevenson University are the only schools surveyed that had more than six digital forensic courses available, with the majority of universities only having between one and four courses. Of course, with such a small selection, it is fair to assume that the large variety of topics that digital forensics encompasses are not being thoroughly examined, but that these courses likely serve as an overview.

Even within their own respective categories, there are still dissimilarities within undergraduate and graduate programs. Undergraduate programs may or may not require prerequisites to their various digital forensics courses, digital forensics may or may not be its own degree program, it may or may not be included in a different existing degree program, and it may or may not be offered as an elective course or courses [27]. Even degree programs can vary, with some focusing on criminal justice and law coursework,

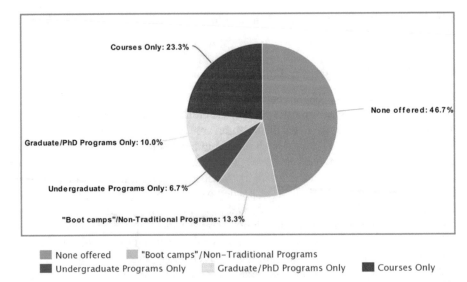

Fig. 1. Digital forensics programs, courses, and concentrations offered by top STEM universities in the US

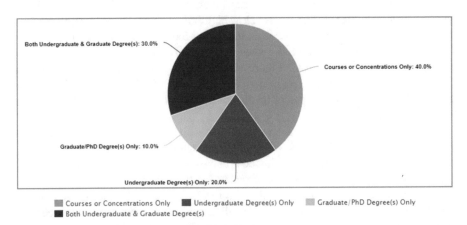

Fig. 2. Breakdown of programs offered at top universities for digital forensics in the US

and others emphasizing the technology and computers coursework. Graduate programs are similar in that they are also dissimilar. Some programs require relevant undergraduate degrees or prerequisite courses, while others do not. Some programs offer courses more similar to traditional forensic sciences, while others offer technology intense courses, while still others offer anything between one overview course to a large selection of electives to pick from. There has traditionally been more room for elective choice at the graduate level [19], which may be favorable to the wide variety of options.

From the universities examined, ninety-two courses offered were examined using the keywords: digital forensics, computer forensics, and forensic/forensics. Figure 3 below shows the distribution of course topics from most frequent to least frequent.

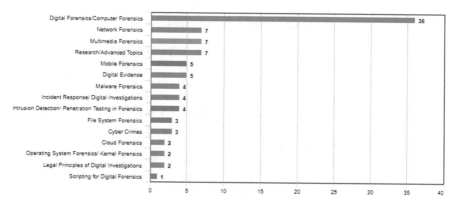

Fig. 3. Distribution of course topics in digital forensic education

3 Discussion

3.1 Understanding Gaps Between Topics Taught and Emerging Trends

As shown in Fig. 3, an overwhelming majority of digital forensics courses serve as an introductory overview-type course. This majority is followed by Network Forensics, Multimedia Forensics, and Research-based Advanced Topics courses, while subfields such as File System Forensics, Mobile Forensics, and Cloud forensics appear further down the chart. Other subfields such as IoT forensics, vehicle forensics, or database forensics did not appear at all, indicating that of the universities surveyed, none of them offered courses related specifically to these subfields. Though IoT is considered a subfield of digital forensics, which in itself is often considered a subfield or niche area of study, IoT is an emerging trend and IoT devices are beginning to have a substantial place in everyday life, which will only increase as technology progresses [22]. IoT devices encompass medical equipment and devices as well as "smart" home systems such as thermostats, door locks, kitchen appliances, sprinkler systems, lightbulbs, and more. IoT Forensics could be used to examine any of these physical world items, and its importance may be more significant due to the immature nature of many IoT devices. IoT forensics tools are not yet refined or widely available, and often IoT devices have completely unique hardware and software due to the sizable number of different devices, and therefore manufacturers of said devices.

Universities and institutions that provide avenues for digital forensics education are certainly making progress in program and course development. However, IoT forensics is just one example of the gap between available topics in digital forensics education and emerging technology trends. According to Harneker and Stander [20], the top three trends are digital forensic processes, image forensics, and cloud forensics. Digital forensics courses tend to be generalized, with 39.1% of all courses surveyed being overviews of the field. In comparison, only 7.6% of courses focus on Network Forensics, the second most offered course. Emerging trends such as mobile device forensics and cloud forensics are covered, though only in a small number of courses. Other trends such as IoT forensics, vehicle forensics, or database forensics do not yet have courses focused on those topics specifically.

3.2   Challenges in Developing Digital Forensic Education Programs

According to research and literature examining the subject of digital forensics and education, the multidisciplinary nature of digital forensics is a significant hurdle in course and program development. It is difficult and perhaps unwise to focus too much on one topic or subfield, because even typical forensic investigations are multi-faced in nature [24]. Some researchers have suggested that the best way to develop a digital forensics curriculum is to look at local employers' needs and design a program that will meet those needs [29]. Others, like Liu [27] suggest that curricula structure should be based on factors such as: analysis of current practices, definitions of pathway missions, objectives, and learning outcomes, analysis of needed levels of knowledge and skills of emerging trends in forensic computing, and course themes including digital investigations, criminal justice, digital forensics, ethical hacking, and digital evidence. Core themes that may encompass the multiple disciplines involved in digital forensics would include the technology side - operating systems, file systems, data storage, and more - as well as the criminal justice and legal matters, including relevant legal systems, legal processes, relevant laws, and the regulatory environment of handling digital evidence and forensic investigations [30]. Still others argue that digital forensics is too broad, and curricula should focus not on the specific disciplines and emerging topics, but instead focus on teaching students how to learn for themselves [33].

Once topics are decided on for a course or a program, even more challenges arise. As expected, there is not an agreed upon method among digital forensics researchers about how to implement teaching practical skills alongside general knowledge and thinking skills. There is, however, at least a general consensus that practical skills are invaluable in a digital forensic setting. Some existing programs are more practical, while others place more emphasis on lectures. Several researchers have pointed out the importance of compiling learning objectives and overall educational goals while designing curricula. Kiper [23] advises the use of Bloom's Taxonomy, which is an educational framework of educational models used to classify learning objectives. Of course, it may still be difficult to design digital forensics courses that cover some of the multidisciplinary topics to students of various backgrounds. For example, students with no background in technology may struggle with advanced technological topics, and students with no background in law or legal matters may struggle with those topics, if they are not introduced at an introductory level. Though some programs require prerequisites, a large percentage of courses are general overviews of digital forensics. It may be difficult to cover topics at a depth that all students will be challenged enough to learn, but not so overwhelmed that learning is impossible. According to Bicak et al. [7], an overloaded curriculum, or curriculum bloating, will likely only cause drop outs and bad grades. It is important to consider the level of risk and complexity of highly specific technology, as well as the turnover times for curricula approval, equipment costs, and software requirements while developing a new digital forensics program to keep up with emerging topics and technologies.

In addition to these challenges, there may be administrative challenges to those who wish to create a digital forensics program or degree plan. For instance, in addition to the lack of standardization, there are also challenges such as lack of textbooks and qualified faculty and educators available. Solutions to these challenges tend to be solutions such as "create textbooks" or "educate professionals" which are certainly goals in advancing digital forensics education as a whole, but do not pose specific, helpful, methodical solutions to either problem. Administrative challenges may also arise due to the fact that digital forensics does not always fall under a similar department or college. As shown in Table 2, several different departments or colleges may accommodate digital forensics programs or courses. Finding faculty in an Engineering or Criminal justice department that are qualified to teach digital forensics may prove more problematic than locating such an individual in a Computer Science or Cybersecurity department, for instance.

3.3 Implications to Policy Makers and Higher Education

Through examining existing digital forensics programs and the challenges that arise in developing new programs, it is obvious that overall, current education is scattered at best. The lack of structure is clear in the industry as it struggles to fill digital investigator positions, and the shortage of professionals with relevant skills hinders incident response [29]. As the scope of the field has increased with a number of subfields as new technologies emerge, digital forensics may be too big to be considered as just a subfield itself of more established technological fields. Gaps in covered materials suggest a complete lack of education in subfields such as IoT, vehicle, and database forensics. Policy makers and developers of higher education must find a better way to combine core and practical knowledge, while also covering recent advances and research in digital forensics as a whole. Relevant areas of study, including the legal field, require some form of accreditation and credentialing, often at the state or even national level [36]. Perhaps it is time to develop similar standards and or at least improve structured education for the field of digital forensics.

Examining the evolution of cybersecurity as an educational field may prove useful in reviewing options for developing digital forensics programs. In fact, in the early days of cybersecurity, curricula used to be just a few specialized courses in a computer science or similar program, much like digital forensics is today [7]. Now, cybersecurity degrees are becoming more common as independent programs. Some researchers suggest using the National Initiative for Cybersecurity Education (NICE) framework, developed by the National Institute of Standards and Technology (NIST), to develop digital forensics curricula. As both cybersecurity and digital forensics evolve, more and more courses involving the words "forensics" and "security" are emerging [3]. Cyber crimes and threats are ever increasing, and cybersecurity, cyber intelligence, and digital forensics are potentially complementary and share some underlying core values. It is certainly plausible that a framework that worked well for the development of cybersecurity education would be promising for the study of digital forensics.

4 Conclusion

In conclusion, educational standards and program development is essential to improve the quality of future and existing digital forensics education. Interest in the field of digital forensics and in relevant subfields is growing, along with the rapidly growing field itself. Programs and courses are being developed, but there is still no solid agreement on standards or specific learning outcomes to be provided by these educational materials. It is important to invest in figuring out the best methods and practices to improve the study of digital forensics sooner rather than later, to educate professionals and keep up with advancing technology.

A Appendix

Information presented in the table below is accurate as of 12 Nov 2020.

University	Course Catalog
California Institute of Technology	http://catalog.caltech.edu
Massachusetts Institute of Technology	https://mitadmissions.org/discover/the-mit-education/majors-minors/
Rice University	https://www.rice.edu/departments#block-departmentsundergraduatemajors https://techbootcamps.rice.edu/cybersecurity/#1599676490363-5bb2bd04-96d7
Michigan Technological University	https://www.mtu.edu/computing/graduate/cybersecurity/ https://www.mtu.edu/computing/undergraduate/cybersecurity/curriculum/
Stevens Institute of Technology	https://www.stevens.edu/sites/stevens_edu/files/Stevens_2020-2021_Academic_Catalog.pdf#page=241
Missouri University of Science and Technology	http://catalog.mst.edu/undergraduate/degreeprogramsandcourses/computerscience/#courseinventory
Georgia Institute of Technology	https://pe.gatech.edu/courses/digital-forensics-for-incident-response https://www.gatech.edu/academics/all-degree-programs
Illinois Institute of Technology	http://bulletin.iit.edu/graduate/colleges/applied-technology/department-information-technology-management/master-cyber-forensics-security/ http://bulletin.iit.edu/undergraduate/colleges/computing/information-technology-management/#coursestext
Carnegie Mellon University	https://www.cmu.edu/ini/academics/cyfir.html
Case Western Reserve University	https://bootcamp.case.edu/cybersecurity/
Purdue University	https://polytechnic.purdue.edu/degrees/phd-technology/specialization-cyber-forensics https://polytechnic.purdue.edu/degrees/cybersecurity

(continued)

(*continued*)

University	Course Catalog
Johns Hopkins University	https://e-catalogue.jhu.edu/engineering/engine ering-professionals/cybersecurity/cybersecurity-master-science/#requirementstext
University of Michigan	http://catalog.umd.umich.edu/archives/2016-2017/ undergraduate/college-engineering-computer-sci ence/digital-forensics/#majortext http://catalog.umd.umich.edu/undergraduate/col lege-engineering-computer-science/cyber-security-information-assurance/#majortext
Tuskegee University	https://www.tuskegee.edu/programs-courses/col leges-schools/cbis/computer-science/graduate-pro gram https://www.tuskegee.edu/programs-courses/col leges-schools/cbis/computer-science/information-technology-major
University of Tulsa	https://bulletin.utulsa.edu/content.php?filter% 5B27%5D=-1&filter%5B29%5D=&filter%5Bc ourse_type%5D=-1&filter%5Bkeyword%5D=for ensics&filter%5B32%5D=1&filter%5Bcpage% 5D=1&cur_cat_oid=26&expand=&navoid=1445& search_database=Filter#acalog_template_course_ filter
Drexel University	http://catalog.drexel.edu/undergraduate/collegeof computingandinformatics/computingandsecurityt echnology/#degreerequirementstext
Florida State University	http://criminology.fsu.edu/degrees/
Keiser University	https://www.keiseruniversity.edu/bachelor-science-cyber-forensicsinformation-security/
Christian Brothers University	http://cbu.smartcatalogiq.com/2020-2021/Catalog/ Policies-Undergraduate-Programs/School-of-Sci ences/Copy-of-Cyber-Security-and-Digital-Forens ics-Bachelor-of-Science
Champlain College	https://online.champlain.edu/degrees-certificates/ masters-digital-forensic-science https://online.champlain.edu/degrees-certificates/ bachelors-computer-forensics-digital-investigations
Robert Morris University	https://www.rmu.edu/academics/undergraduate/cyb ersecurity-bs
George Mason University	https://catalog.gmu.edu/colleges-schools/engine ering/electrical-computer/digital-forensics-ms/#req uirementstext

(*continued*)

(continued)

University	Course Catalog
Stevenson University	https://www.stevenson.edu/academics/undergraduate-programs/cybersecurity-digital-forensics/ https://www.stevenson.edu/online/academics/online-graduate-programs/cybersecurity-digital-forensics/ https://www.stevenson.edu/online/academics/online-certificate-programs/digital-forensics/
Strayer University	https://www.strayer.edu/online-degrees/bachelors/criminal-justice/computer-forensics https://www.strayer.edu/online-degrees/masters/information-systems/computer-forensic-management https://strayer.smartcatalogiq.com/en/2015-2016/Catalog/Programs/University-Minors/Computer-Forensic-Management-Minor
Utica College	https://programs.online.utica.edu/programs/masters-cybersecurity/curriculum?cmgfrm=https%3A%2F%2Fwww.google.com%2F https://programs.online.utica.edu/programs/bachelors-cybersecurity/curriculum?cmgfrm=https%3A%2F%2Fwww.google.com%2F
University of Arizona	https://www.arizona.edu/degree-search/majors/cyber-operations-defense-and-forensics-emphasis

References

1. 2021 Best Colleges with Computer Forensics and Counterterrorism Degrees (2020). https://www.niche.com/colleges/search/best-colleges-with-cyber-computer-forensics-and-counterterrorism/. Accessed 12 Nov 2020
2. Best Schools for STEM Majors, 26 May 2020. https://www.bachelorsdegreecenter.org/best-schools-stem-majors/. Accessed 12 Nov 2020
3. Antunes, M., Rabadão, C.: Cybersecurity and digital forensics – course development in a higher education institution. In: Madureira, A.M., Abraham, A., Gandhi, N., Silva, C., Antunes, M. (eds.) SoCPaR 2018. AISC, vol. 942, pp. 338–348. Springer, Cham (2020). https://doi.org/10.1007/978-3-030-17065-3_34
4. Bashir, M., Campbell, R.: Developing a standardized and multidisciplinary curriculum for digital forensics education (2019)
5. Belshaw, S.H.: Next generation of evidence collecting: the need for digital forensics in criminal justice education. J. Cybersecur. Educ. Res. Pract. (2019). https://digitalcommons.kennesaw.edu/jcerp/vol2019/iss1/3/
6. Best Colleges with Cyber/computer Forensics and Counterterrorism Degrees in the U.S. (2020). https://www.universities.com/programs/cyber-computer-forensics-and-counterterrorism-degrees. Accessed 12 Nov 2020
7. Bicak, A., Liu, M., Murphy, D.: Cybersecurty curriculum development: introducing specialties in a graduate program. Inf. Syst. Educ. J. (ISEDJ) **13**(3), 99–110 (2015)

8. The Carnegie Classification of Institutions of Higher Education ® (2018). https://carnegiec lassifications.iu.edu/classification_descriptions/basic.php. Accessed 12 Nov 2020
9. Clapper, J.R.: Worldwide threat assessment of the US intelligence community, 29 January 2014. https://www.dni.gov/files/documents/Intelligence%20Reports/2014%20W WTA%20%20SFR_SSCI_29_Jan.pdf
10. Coudriet, C.: Top 25 STEM Colleges 2018, 21 August 2018. https://www.forbes.com/sites/ cartercoudriet/2018/08/20/top-25-stem-colleges-2018/?sh=71463ab31f8b. Accessed 12 Nov 2020
11. Crimson Education US - 12 Top STEM Universities in the US, 1 August 2018. https://www.cri msoneducation.org/us/blog/campus-life-more/top-stem-universities/. Accessed 12 Nov 2020
12. Dafoulas, G.A., Neilson, D.: An overview of digital forensics education. In: 2019 2nd International Conference on New Trends in Computing Sciences (ICTCS) (2019)
13. Delija, D., et al.: Implementation of virtual digital forensic class and laboratory for training, education and investigation. In: MIPRO 2019, pp. 1175-1180 (2019)
14. Deshpande, P., Ahmed, I.: Topological scoring of concept maps for cybersecurity education. In: SIGCSE 2019, pp. 731–737 (2019)
15. Digest of Education Statistics (2019). https://nces.ed.gov/programs/digest/d19/tables/ dt19_322.10.asp?current=yes. Accessed 12 Nov 2020
16. Doherty, E., Laubsch, P., Goei, E.: An example of project based learning to advance women's interest in STEM education and robotics. In: 2019 International Conference on Computational Science and Computational Intelligence (CSCI), pp. 763–767 (2019)
17. Drange, T., Irons, A., Drange, K.: Creativity in the digital forensics curriculum. In: Proceedings of the 9th International Conference on Computer Supported Education (CSEDU 2017), vol. 2, pp. 103–108 (2017)
18. George, J.C.: Examining the sameness of postsecondary digital forensics curricula: a content analysis. Doctoral dissertation, Northcentral University, ProQuest LLC, Ann Arbor, MI (2020)
19. Gottschalk, L., Liu, J., Dathan, B., Fitzgerald, S., Stein, M.: Computer forensics programs in higher education: a preliminary study. In: SIGCSE 2005: Proceedings of the 36th SIGCSE Technical Symposium on Computer Science Education, February 2005, pp. 147–151 (2005)
20. Harneker, R., Stander, A.: Developing a digital forensics curriculum: exploring trends from 2007 to 2017. In: Tait, B., Kroeze, J., Gruner, S. (eds.) SACLA 2019. CCIS, vol. 1136, pp. 64–76. Springer, Cham (2020). https://doi.org/10.1007/978-3-030-35629-3_5
21. Hasan, R., Zheng, Y., Walker, J.T.: Digital forensics education modules for judicial officials. In: Choo, K.-K.R., Morris, T., Peterson, G.L., Imsand, E. (eds.) NCS 2020. AISC, vol. 1271, pp. 46–60. Springer, Cham (2021). https://doi.org/10.1007/978-3-030-58703-1_3
22. Karabacak, B., Aydin, K., Igonor, A.: IOT forensics curriculum: is it a myth or reality? In: Annual ADFSL Conference on Digital Forensics, Security and Law (2019)
23. Kiper, J.: Forensication education: towards a digital forensics instructional framework. SANS Institute Information Security Reading Room (2017)
24. Leung, W., Blauw, F.F.: An augmented reality approach to delivering a connected digital forensics training experience. Inf. Sci. Appl. **621**, 350–361 (2019)
25. Liu, J.: Developing an innovative baccalaureate program in computer forensics. In: 36th ASEE/IEEE Frontiers in Education Conference, s1h (2006)
26. Liu, J.: Baccalaureate programs in computer forensics. In: 2016 IEEE International Conference on Electro Information Technology (EIT), pp. 0615–0620 (2016)
27. Liu, J.: Ten-year synthesis review: a baccalaureate program in computer forensics. In: SIGITE 2016: Proceedings of the 17th Annual Conference on Information Technology Education, pp. 121–126 (2016)
28. Luciano, L., Baggili, I., Topor, M., Casey, P., Breitinger, F.: Digital forensics in the next five years. In: ARES 2018 (2018)

29. Naqvi, S., Sommer, P., Josephs, M.: A research-led practice-driven digital forensic curriculum to train next generation of cyber firefighters. In: 2019 IEEE Global Engineering Education Conference (EDUCON), pp. 1204–1211 (2019)
30. Palmer, I., Wood, E., Nagy, S., Garcia, G., Bashir, M., Campbell, R.: Digital forensics education: a multidisciplinary curriculum model. In: James, J.I., Breitinger, F. (eds.) ICDF2C 2015. LNICSSITE, vol. 157, pp. 3–15. Springer, Cham (2015). https://doi.org/10.1007/978-3-319-25512-5_1
31. Roy, S., Wu, Y., Lavenia, K.N.: Experience of incorporating NIST standards in a digital forensics curricula*. In: 2019 7th International Symposium on Digital Forensics and Security (ISDFS) (2019). https://doi.org/10.1109/isdfs.2019.8757533
32. Seda, M.A., Kramer, B.P., Crumbley, D.: An examination of computer forensics and related certifications in the accounting curriculum. J. Digit. Forens. Secur. Law **14** (2019)
33. Tang, L.: The construction of electronic examination for the course of information crime and computer forensics. In: 5th International Conference on Social Science and Higher Education (ICSSHE 2019). Advances in Social Science, Education and Humanities Research, vol. 336, 734–737 (2019)
34. Verma, R., Bansal, P.: Scope of managing knowledge in digital forensics. In: International Conference on Sustainable Computing in Science, Technology & Management (SUSCOM-2019), pp. 2381–2387 (2019)
35. Wu, T., Breitinger, F., Baggili, I.: IoT ignorance is digital forensics research bliss: a survey to understand IoT forensics definitions, challenges and future research directions. In: ARES 2019 (2019). https://doi.org/10.1145/3339252.3340504
36. Zahadat, N.: Digital forensics, a need for credentials and standards. J. Digit. Forens. Secur. Law **14** (2019)

Designing a Cybersecurity Curriculum Library: Best Practices from Digital Library Research

Blair Taylor[1]([⊠]), Sidd Kaza[1], and Melissa Dark[2]

[1] Towson University, Towson, MD, USA
btaylor@towson.edu
[2] Dark Enterprises, Lafayette, IN, USA

Abstract. The global cybersecurity crisis has forced academic institutions to expand cybersecurity education, and accessible, quality cybersecurity curriculum is needed. The CLARK cybersecurity curriculum resource library is a large-scale Design Science Research project intended to address critical demands in cybersecurity. This paper focuses on the need for a living digital library, surveys current cybersecurity repositories, and discusses factors that determine the success of a digital library. We also discuss CLARK, a multifaceted solution that includes various lenses for building and sustaining a living library of cybersecurity curriculum.

Keywords: Cybersecurity education · Cybersecurity curriculum · Cybersecurity resources · Digital library

1 The Need for a Cybersecurity Digital Library

Cybersecurity has critical implications for our nation's economic, social, information, military, and physical infrastructure. This has challenged academic institutions to transition research into education as well as expand the cyber workforce. To this end, various funding agencies have supported projects that develop cybersecurity curricular resources. An innovative approach to leverage this curriculum is needed to transform cybersecurity education and research. This problem requires designing a system for knowledge management, information retrieval, and usage analysis. The design process involves studies dealing with the evaluation, adaption, use, and effectiveness of a library of cybersecurity resources. Academic institutions have their own unique issues with cultural inertia and technology acceptance; appropriate research methodologies can be used to examine perceived usefulness impact, valuation, and management of the library artifact across diverse institutions.

The goal of design science research (DSR) is to build and design an innovative product or artifact to address field problems using theory-based research (Peffers et al. 2007). The DSR lifecycle requires that researchers identify a relevant problem, understand and expand appropriate research theory, and perform continuous and rigorous evaluation. In this article, we present the CLARK project at Towson University within the context of DSR frameworks. We describe in detail the DSR activities carried out (problem

K.-K. R. Choo et al. (Eds.): NCS 2021, LNNS 310, pp. 47–59, 2022.
https://doi.org/10.1007/978-3-030-84614-5_5

diagnosis, technology invention) and the lessons learned that might benefit similar DSR projects.

Cybersecurity is a critical global issue, and a skilled and knowledgeable workforce is crucial to the continued national and economic security of the nation. While collegiate cybersecurity programs have grown dramatically over the past 20 years, the demand for cybersecurity professionals is increasingly outpacing the supply. It is vital that more colleges establish and expand cybersecurity programs in order to meet labor demand. Concurrently, there is an acutely growing need to develop K-12 cybersecurity courses that produce students who matriculate into collegiate programs of study. Given the rapidly evolving nature of cybersecurity, faculty need high-quality and up-to-date curriculum along with available, robust, and secure infrastructure. Cybersecurity is a desired common good, i.e., beneficial for all. In order to mobilize the educational sector to provision this good, we need a common pool of educational resources.

Digital libraries have been around for over twenty-five years. They are logical extensions of physical libraries that do the following: 1) amplify existing resources and services, 2) enable new resources and services, and 3) allow extended access (especially important in an increasingly virtual world). A motivating philosophy of many digital library (DL) efforts is to afford more access to more relevant information in a more expedient and cost-effective manner. And just as physical libraries have long served the important role of bringing information to *all*, digital libraries are impelled by the same aspiration to bridge digital divides. Digital libraries can house collections of digital works and/or collect pointers to other resources and thus provide a single point of access to a wide range of autonomously distributed resources (Jeng 2005). Digital libraries are enabling technologies for digital asset management in the realms of electronic publishing, teaching and learning, research, and other activities. Some digital libraries focus on the archiving and sharing of educational resources, and the observations included here focus specifically on educational resources digital libraries.

We already have early innovators responding to the need for curriculum sharing in cybersecurity education, such as CyberWatch (*Library - National CyberWatch Center: National CyberWatch Center* n.d.), Department of Homeland Security (DHS), and SkillsCommons.org (*Home - SkillsCommons Repository* n.d.) There are similar efforts in computer science (CS), such as Ensemble (Shipman et al. 2010) and EngageCSEdu (*Search Instructional Materials—EngageCSEdu* n.d.), and in other Science, Technology, Engineering, and Mathematics (STEM) fields as well. The existing repositories offer several good features and a solid base on which to build. Table 1 shows a comparative evaluation of several representative repositories.

Ensemble. (Shipman et al. 2010), funded by the National Science Foundation (NSF), includes content on computer science, computer engineering, software engineering, information science, information systems, and information technology, and a small, featured collection on security. *Submission mechanism* - user registration is required with limited curation of curriculum.

Carnegie Mellon University, Software Engineering Institute (CERT). (*Software Assurance Materials and Artifacts | The CERT Division,* n.d.) This repository hosts materials on several topics in software assurance. *Submission mechanism* - The content is a curated set, consisting of courses taught at Carnegie Mellon and other universities.

Table 1. Digital library comparison

Digital library	Focus of curriculum	Cybersecurity category?
Ensemble	Computer Science	Yes
CERT (Software Engineering Institute)	Software Assurance	Yes
National CyberWatch Center	Cybersecurity	N/A
SkillsCommons.org	Workforce training	Yes
National Science Digital Library	Science/Tech./Eng./Math	No
Computer Science Teachers Association (CSTA) K-12 Repository	Computer Science	No
EngageCSEdu	Computer Science, CS1, CS2	No

National CyberWatch Center. These cybersecurity resources are searchable by education level, job classification, and type (materials, reports, etc.). *Submission mechanism* - Educators can contribute their curriculum resources to the repository by registering on the portal. There is limited peer review of resources.

Skillscommons. Skillscommons includes free and open educational resources in workforce development. It is primarily focused on training materials at the community college/vocational school level. *Submission mechanism* - The material in Skillscommons is limited to the recipients of TAACCCT grants (https://doleta.gov/taaccct/). There appears to be limited peer review of resources at submission to the repository.

National Science Digital Library (NSDL). (Curated Collections | NSDL, n.d.) The NSDL provides online educational resources for STEM disciplines. The collection in this repository includes links to a wide variety of resources hosted on other sites. *Submission mechanism* - Educators can contribute their curriculum resources to the repository by registering to the portal and providing the resource URL. There is limited review of resources.

Computer Science Teachers Association (CSTA) Source Online K-12 Repository. (CSTA Source Online K-12 Repository | CSTA Source, n.d.) CSTA Source holds K-12 materials peer-reviewed by CSTA. It is only open to CSTA members. It is organized by CSTA Computer Science standards. *Submission mechanism* - The submission is open and is available without authentication.

EngageCSEdu. The materials in EngageCSEdu are limited to introductory CS courses (*Search Instructional Materials | EngageCSEdu*, n.d.). At the time of writing, the repository contained over 1,500 unique instructional materials that cover all major CS1 and CS2 programming languages and over 325 introductory computer science topics. *Submission mechanism* - EngageCSEdu allows faculty to contribute new materials with logins.

2   What Makes a Successful Digital Library?

Given that digital libraries have existed for over two decades and that many have failed, we focus on characteristics that are common to successful, sustainable digital libraries. Figure 1 shows the lifecycle for a successful digital library through inception, creation, growth, and maturity (Mullins et al. 2002).

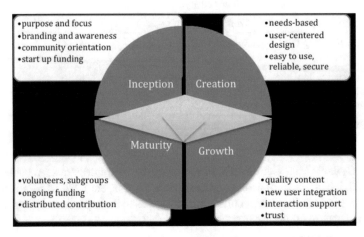

Fig. 1. What makes a successful digital library - using a lifecycle model (Mullins et al. 2002)

Several characteristics were found to be common to successful digital libraries during inception (Calhoun 2014). Successful digital libraries have grown *out of* the communities they intended to serve, based on a purpose(s) articulated within that community. During inception, the effort needs to have alignment and focus around a clear, compelling mission. As the digital library evolves from idea to entity, the mission can evolve, however clarity of purpose remains critical to success. Starting with inception but continuing through the other phases in the life cycle, it is important to test and validate assumptions about need/purpose, appeal of content, characteristics, expectations, and work practices of target users. Because the digital library has no branding of its own before/during inception, research on successful digital libraries shows that they initially draw their branding through the reputation(s) of visionary, determined leadership who are highly visible and credible. However, the effort must quickly establish its own brand, and the brand must communicate the identity, intent, and nature of the digital library in a way that resonates with the target audience and appeals to the target audience's core values.

In addition to branding what the digital library is/can do, (Marchionini et al. 2003) investigations have found it is also important to brand what the digital library is not/cannot do. Successful digital libraries have been found to use a variety of practical techniques to establish branding and awareness including: 1) associating the new digital library with a destination site that is recognized and well-respected by target audiences, 2) emphasizing unique features (if they exist), 3) making it highly visible through search engines, and 4) making it open access (Calhoun 2014).

Research on successful digital libraries shows that during creation and growth, there is growing emphasis on the concept of the digital library as first and foremost a social space and not its technical functionality. This necessitates understanding human information needs and the behaviors/tasks that arise from those needs, as well as considering how the digital library affects subsequent human information needs and behaviors (Marchionini et al. 2003). During creation, it is critical to use the information needs, characteristics, and contexts of people who will or may use the library in each stage of design, implementation and evaluation. The challenge is to create a virtuous circle, where all stakeholders are involved in a manner that encourages the actions that maintain and improve the resource.

Marchionini et al. (2003) studied three digital libraries and identified several important characteristics during the creation and growth phases of digital libraries. Common to digital libraries that have succeeded is a framework that characterizes the granularity, size and nature of objects in the digital library, and communicates those to stakeholders efficiently. Successful digital libraries know the critical importance of discoverability, i.e., helping users find educational content that takes account of their context. Unfortunately, authors of open educational resources are notoriously negligent when it comes to filling out metadata fields, and with many open educational resources (OERs), there is no one else to do that cataloging (Kortemeyer 2013). Kortemeyer (2013) suggests that tighter coupling of the repository with the deployment of it (usually through a content management system) could help gather such dynamic metadata. One idea for handling this challenge is to create value for authors to incentivize them to enter the metadata. This addresses part of the challenge; users would be able to discover content based on what the author "projects" to be the intended usage, which is not as good as the suggestion of dynamic metadata based on usage presented in (Kortemeyer 2013) but is better than nothing. Successful digital libraries support multiple formats and have established efficient means of handling digital rights managements. During creation and growth, users will expect a quality collection. There are many dimensions of quality to be considered with regard to content, such as: accuracy, accessibility, comprehensiveness, clarity, currency, relevancy, consistency, and efficacy (Zhang 2010). Beyond a quality collection, successful digital library efforts show that users also expect a sizable collection. Value is provided to users by providing ample quality content and enhanced by helping users make connections across different items in the collection. Marchionini et al. (2003) also found that users benefit from being pointed to other digital libraries when the current repository does not have what the user is seeking; this is the value of the digital library as a networked resource.

Usability and usefulness are paramount. One of the assets of the digital library is the 24-7-365 access that allow users to work at their convenience. However, this characteristic makes it important to create a digital library that supports users without human intervention as much as possible. Usability refers to functions such as "is it available to me?", "can I click on that link?", "can I easily contribute a resource?", "can I easily search and find what I need?", etc. (Jeng 2005). The barriers to getting started need to be low and in close alignment with how the target audience works. Usefulness refers to such functions as "did this help me?", "was it worth the effort?" (Jeng 2005). In addition, the system must be stable, reliable, have adequate performance, have effective access rights, and ensure security and privacy.

The last life cycle phase is maturity. The Achilles heels of successful digital libraries are: 1) the financial sustainability, and 2) clarity about who has ongoing responsibility for the DL (Calhoun 2014). Frequently start-up funding is exhausted by this time, and while leaders still need to identify a clear value proposition, it is imperative that direct costs are minimized, and funding sources are diversified. Possible funding mechanisms for digital libraries are listed in Table 2.

Table 2. Possible funding mechanisms for digital libraries

Revenue	Nonfinancial support
Memberships/subscriptions	Volunteer labor
Content licensing	Partnerships
Advertising	Support from host institution
Scholarships	Other in-kind support (free rent, tech support,
Endowment income	free server space, waived F&A)
Grants	Controlling and reducing costs
Sponsorship	
Government subsidy	
Open access author pays	
Premium services (combined with freely	
available ones)	
Hybrids of the above	

3   CLARK - Cybersecurity Curriculum Library

CLARK started at Towson University with funding from the National Security Agency (NSA) to build and share cybersecurity resources and subsequently broadened in scope to develop methodologies for designing, collecting, searching, and analyzing effective cybersecurity learning objects. The CLARK system, hosted at www.clark.center, was created as a high-availability, relevant, and scalable repository for curricular resources in the cybersecurity education community. The existing repositories (described above) offer several good features and a foundation on which to build, however, there is a clear need for a model that focuses solely on quality cybersecurity curriculum (Kortemeyer 2013; Zhang 2010). The design and implementation of the CLARK system has been informed by research in the area of digital libraries. The strength of CLARK lies in three distinct components: its evidence-based model for developing quality cybersecurity curriculum; the CLARK system; and its unique collections, described in detail below.

3.1   The CLARK Curriculum Model

The criticality of the cybersecurity crisis requires high-impact solutions. The CLARK curriculum model focuses on developing quality content: curriculum that is technically correct and relevant while being instructionally sound, usable and adoptable. The curriculum development process adopted by CLARK follows the system development lifecycle analogy, described below.

Requirements. The CLARK curriculum development lifecycle begins with requirements analysis to identify crucial cybersecurity areas where curriculum is needed. This may be guided by workforce frameworks, curriculum guidelines, specific job needs of industry and government, or programmatic needs in academia. The current collections in CLARK were created based on a gap analysis by funding agencies, including NSA's National Cybersecurity Curriculum Program (NCCP), curriculum guidelines such as CS 2013 (e.g., Security Injections collection focused on CS 2013[1] IAS area) and CSEC 2017, and the need to create a nationwide network of community colleges that have met national standards in cybersecurity education (the C5 collection[2]).

Design. The CLARK design process begins with the development of meaningful learning outcomes that identify the appropriate level of student attainment using Bloom's taxonomy. Faculty convert their curriculum to standardized templates, categorize it as a nanomodule, module, etc. according to completion time (see Fig. 3) and organize it in the appropriate hierarchical structure. Further categories include collection type, academic level (elementary, middle, undergraduate, etc.), and mapping according to curricular guidelines. Templates for all curriculum learning objects, nanomodules, lectures, labs, courses, etc., have been developed by instructional designers with the intent of supporting effective and efficient searchability and selection for CLARK users.

Implementation. The implementation cycle of the CLARK model includes an iterative process that takes place between the curriculum author and instructional curator. The instructional curator performs a preliminary review, using evidenced-based practices for effective instruction. The review uses a standardized approach to ensure that content is accurate and current, learning outcomes are appropriate, instruction content and assessment is appropriately aligned, and curriculum is engaging (i.e., it uses active learning and is usable across different classroom settings). Curriculum authors are given feedback and improvements are made based on the preliminary review.

Testing and Release. Once learning objects have been revised, they are submitted to CLARK for testing. Testing is comprised of formal peer review. The review team includes at least two subject matter experts (SME). SMEs conduct a formal review, using a standardized rubric, of each set of learning objects. Again, authors work with the review team to revise curriculum for publication on CLARK. There are many dimensions of quality to be considered with regard to content, such as: accuracy, accessibility, comprehensiveness, clarity, currency, relevancy, consistency, and efficacy (Zhang 2010). This is an expensive process, in terms of time and effort; however, it is needed to ensure that only quality curriculum is published on CLARK to maintain consumer trust. Once curriculum has met quality standards, it is released, i.e., "published", within CLARK.

---

[1] ACM/IEEE-CS Joint Task Force on Computing Curricula 2013. Computer Science Curricula 2013.

[2] Catalyzing Computing and Cybersecurity in Community Colleges (C5).

Maintenance and Relevancy. The challenge of maintaining relevancy in the fast-moving field of cybersecurity requires a formal process. Future work includes designing and implementing a formal relevancy process to ensure that all learning objects in CLARK are timestamped, reviewed, and updated to ensure continued quality.

3.2 The CLARK System

Research shows that successful, sustainable digital libraries are based on known needs and grounded in user-center design content (Marchionini et al. 2003). CLARK was built to address a critical demand to improve cybersecurity education and will expand access to relevant curriculum to assist schools in rapidly provisioning their cybersecurity courses and programs.

The CLARK team utilized an agile methodology to identify user requirements. The team has engaged the community in requirements gathering sessions at three NSF sponsored cybersecurity education workshops (2016–19), CISSE 2018 (Dark et al. 2018) and NACE 2018 (Spafford 2018). Additionally, during development, the CLARK system was thoroughly reviewed by a team of faculty contributors to provide project assessment. Features of the CLARK system, detailed below, include: a Bloom's taxonomy based submission, mapping capability, modularization, and a faceted search.

A Bloom's Taxonomy-Based Submission. The alignment of outcomes, assessment and instruction is foundational to effective curriculum (National Research Council 2001; Wiggins 1999). Building meaningful outcomes is a critical first step in creating appropriate instruction and authentic assessments. When instruction is aligned with outcomes, students are more likely to learn because instructional strategies are selected to invoke the depth of learning being targeted. When assessment is aligned with outcomes, it provides meaningful feedback on whether learning targets were attained.

Most instructors in higher education lack a formal background in instructional design and many struggle with creating effective learning outcomes. The process of creating learning outcomes intends to engage the instructors in thinking beyond what they want to *teach*, to formalizing what they want their students to *learn*. The CLARK system facilitates this process. Its unique entry system including an outcomes builder functionality that provides verb suggestions based on the level of Bloom's learning taxonomy (Fig. 2). This innovative approach sets CLARK apart from other curriculum digital libraries and feedback from users indicates that this improves the quality of curriculum submitted.

Mapping to Curricular Guidelines. Successful digital libraries include comprehensive metadata and help authors catalogue their own content (Marchionini et al. 2003). Incomplete metadata is a weakness of many open educational resources (OERs), authors may neglect to fill out metadata fields and generally, there is no one else to do that cataloging. Kortemeyer (2013) suggests that tighter coupling of the repository with the system could help gather such dynamic metadata. Given the many existing bodies of knowledge, curriculum guidelines, and workforce frameworks; institutions are grappling with ways to find curriculum to build or extend their cyber programs. CLARK includes a recommender system (Fig. 2) that suggests mappings to curricular guidelines including the ACM Computer Science (CS2013)[3], ACM Cybersecurity Guidelines (CSEC2017)[4], NSA CAE Knowledge Units, NICE Workforce Framework, K-12 standards, and other national and global standards.

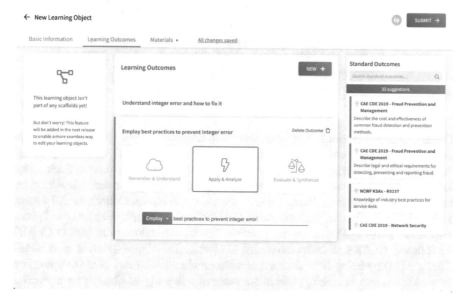

Fig. 2. Learning outcomes builder and mapping interface

Modular Content. Successful digital libraries (Marchionini et al. 2003) include a framework that characterizes the granularity, size, and nature of objects and communicates those to stakeholders efficiently. All curriculum in CLARK is categorized using the taxonomy outlined in Fig. 3. This makes the learning modules easier to quantify and use in the classroom.

Search. Successful digital libraries provide discoverability – more metadata, more cataloguing (Marchionini et al. 2003). Time-strapped faculty want quality curriculum that is quick to find and easy to use. A system should be able to provide the most efficient access for a user, and efficient access will vary by use-case. For instance, faculty exclusively teaching CS1 and CS2 classes may choose to search for cybersecurity curriculum based on courses. Another use-case may be faculty hoping to find curriculum to satisfy a certain CAE knowledge unit requirement. Still another scenario involves faculty looking for modules on "confidentiality" or "software assurance" – topics that naturally span many classes. The CLARK search (Fig. 4) provides flexible and usable interfaces for these users.

---

[3] ACM/IEEE-CS Joint Task Force on Computing Curricula. 2013. Computer Science Curricula 2013. ACM Press and IEEE Computer Society Press. https://doi.org/10.1145/2534860

[4] Curriculum Guidelines for Post-Secondary Degree Programs in Cybersecurity (CSEC 2017).

Learning object	Completion time
Nano-module	Less than 1 hour
Micro-module	1 to 4 hours
Module	Between 4 hours and two weeks (6 hours)
Unit	Between 2 weeks and 15 weeks (45 hours)
Course	15 weeks (45 hours)

Fig. 3. Learning object taxonomy

The CLARK Collections. Learning objects in CLARK are grouped into collections. Each collection implements the CLARK curriculum model slightly differently, though they all go through the design, implementation, testing, and release phases. Each collection has an identified curator who is responsible for interacting with the review team and the technical editing team during various phases on the model. Collections help define areas and communities of practice within CLARK, to allow users to identify and potentially assist in the quality assurance process. Not all curriculum submitted to CLARK is published. CLARK includes references to other digital libraries, so that users benefit (Zhang 2010) by being from directed elsewhere when the current repository does not have what the user is seeking; this is the value of a free digital library as a networked resource.

4   Future Directions

4.1   Outreach and Innovation

Cybersecurity research and its transition to teaching is an essential part of the cybersecurity education solution. CLARK can be a resource to host cutting-edge materials based on current research. However, it is not enough to build curriculum. As the library matures, there must be regular, national and regional workshops to encourage faculty to leverage the curriculum in CLARK and other libraries.

4.2   Cyber Range

An essential component for both faculty and students is a virtual environment/infrastructure to support labs, competitions, and other work in a secure environment. Given the wide variety of platforms required across curriculum in CLARK, this could be a distributed solution, that uses and extends various cyber ranges and commercial cloud environments across the country.

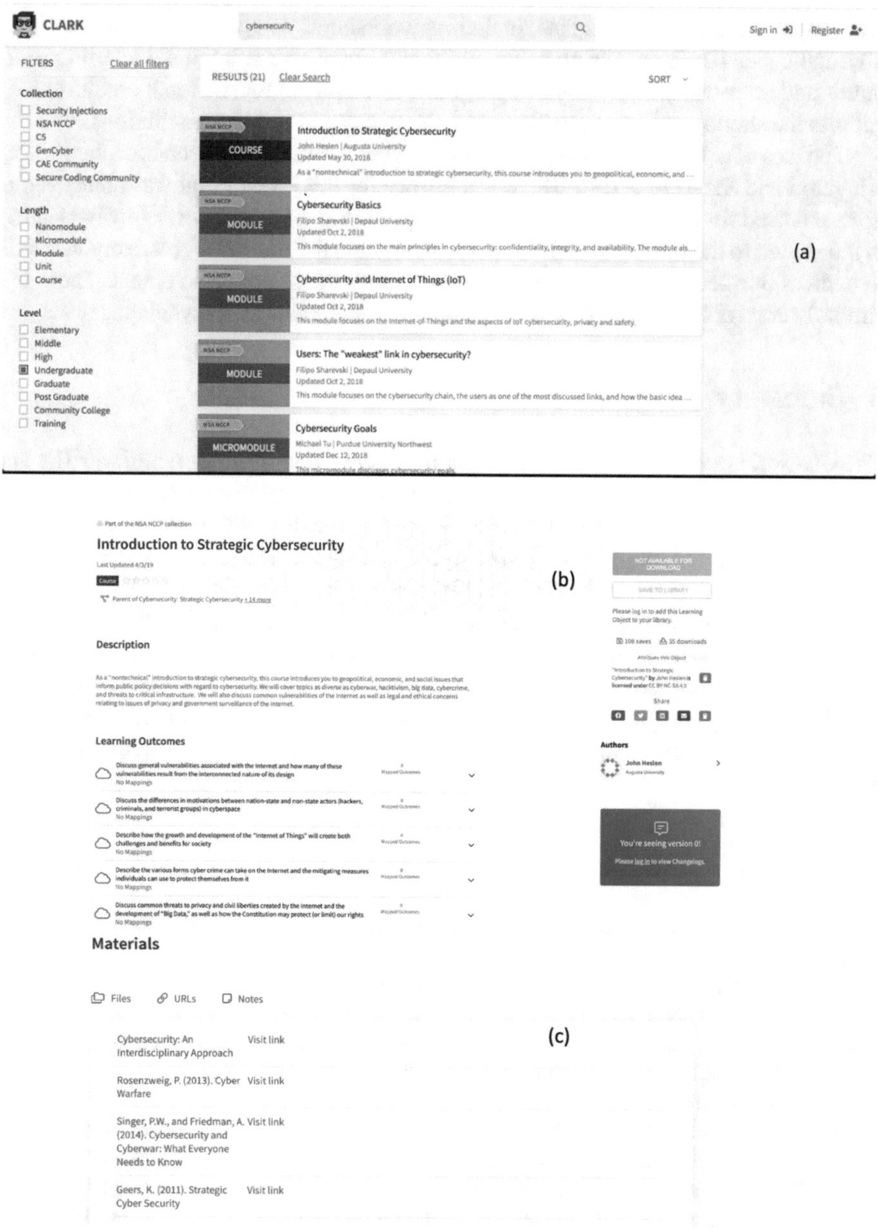

Fig. 4. Faceted search in CLARK – (a) search listing filtered by learning object length, academic level, and collections (b) learning object detail view and (c) materials view

4.3 National Cyber Academy

Cybersecurity curriculum is one component of a comprehensive solution to provide cyber education for the national workforce. Curriculum on CLARK, a cyber environment, and supporting outreach activities, including workshops, comprise the foundations for a virtual cyber academy. We envision additional innovative solutions, including more multi- and inter-disciplinary courses/programs and a model for 'national credit' to help address the shortage of cyber faculty and cyber resources at many institutions.

The idea of a 'bricks and mortar' National Cyber Academy has been broached before (Hagerott and Stavridis 2017). There is a history for this. West Point was conceived to answer a need for land defense leaders and engineering. The need for defensive leaders in the sea led to the United States Naval Academy, and the Air Force Academy was built to address our needs in aerospace, including missiles and atomic weapons. There is a similarly crucial need today educate the Nation's future leaders in cyberspace.

5 Summary

The CLARK system was created based on a demonstrated need for a high-quality and high-availability repository for curricular resources in the cybersecurity education community. With over 3,000 users, CLARK is gaining traction within this community as the go-to source for free peer-reviewed cybersecurity curriculum. Designed based on lessons-learned from research on digital libraries and learning sciences, CLARK is easy to use by both contributors and consumers. In contrast to other digital libraries, which often include barriers of entry that deter contribution, CLARK's unique entry system facilitates curriculum creators in the design of effective learning outcomes and instructional and assessment strategies. Initially developed with NSA funding, CLARK will continue to provide free quality-assured cybersecurity curriculum to schools across the nation.

References

Calhoun, K.: Exploring Digital Libraries: Foundations, Practice, Prospects. Facet Publishing, Abingdon (2014)

CSTA Source Online K-12 Repository | CSTA Source (n.d.)

Curated Collections | NSDL (n.d.)

Dark, M., Kaza, S., LaFountain, S., Taylor, B.: The cyber cube: a multifaceted approach for a living cybersecurity curriculum library. In: The Colloquium for Information Systems Security Education (CISSE) (2018)

Hagerott, M., Stavridis, J.: Home – article – Trump's big defense … – foreign policy guide. Foreign Policy Guide (2017). https://fpguide.foreignpolicy.com/article-trumps-big-defense/

Home - SkillsCommons Repository (n.d.). http://www.skillscommons.org/. Accessed 27 Jan 2021

Jeng, J.: What is usability in the context of the digital library and how can it be measured? Inf. Technol. Libr. 24(2), 47–56 (2005). https://doi.org/10.6017/ital.v24i2.3365

Kortemeyer, G.: Ten years later: why open educational resources have not noticeably affected higher education, and why we should care. Educ. Rev. 1–8 (2013)

Library - National CyberWatch Center: National CyberWatch Center (n.d.)

Marchionini, G., Plaisant, C., Komlodi, A.: The people in digital libraries: multifaceted approaches to assessing needs and impact. In: Digital Library Use: Social Practice in Design and Evaluation. MIT Press, Cambridge (2003). https://doi.org/10.7551/mitpress/2424.003.0009

Mullins, P., et al.: Panel on integrating security concepts into existing computer courses. In: Proceedings of the 33rd SIGCSE Technical Symposium on Computer Science Education - SIGCSE 2002, vol. 34, no. 1, p. 365 (2002). https://doi.org/10.1145/563340.563480

National Research Council: Knowing what students know: the science and design of educational assessment. Issues Sci. Technol. **19**, 48–52 (2001)

Peffers, K., Tuunanen, T., Rothenberger, M.A., Chatterjee, S.: A design science research methodology for information systems research. J. Manag. Inf. Syst. **24**(3), 45–77 (2007). https://doi.org/10.2753/MIS0742-1222240302

Search Instructional Materials | EngageCSEdu (n.d.)

Shipman, F.M., et al.: Ensemble: a distributed portal for the distributed community of computing education. In: Lalmas, M., Jose, J., Rauber, A., Sebastiani, F., Frommholz, I. (eds.) ECDL 2010. LNCS, vol. 6273, pp. 506–509. Springer, Heidelberg (2010). https://doi.org/10.1007/978-3-642-15464-5_68

Software Assurance Materials and Artifacts | The CERT Division (n.d.)

Spafford, E.: New Approaches to Cybersecurity Education (NACE) Workshop (2018). https://www.cerias.purdue.edu/site/nace/

Wiggins, G.: Educative Assessment: Designing Assessments to Inform and Improve Student Performance. Jossey-Bass, San Francisco (1999). https://doi.org/10.5860/choice.36-2887

Zhang, Y.: Developing a holistic model for digital library evaluation. J. Am. Soc. Inf. Sci. Technol. **61**(1), 88–110 (2010). https://doi.org/10.1002/asi.21220

Design of a Virtual Cybersecurity Escape Room

Tania Williams[1]([✉]) [iD] and Omar El-Gayar[2] [iD]

[1] The University of Alabama in Huntsville, Huntsville, AL 35899, USA
tania.williams@uah.edu
[2] Dakota State University, Madison, SD, USA
Omar.El-Gayar@dsu.edu

Abstract. As cybersecurity education and training methods endeavor to build a skilled workforce to meet the growing demands of government and industry, it is important to craft training exercises to build and assess a learner's understanding of cybersecurity topics in a fun and meaningful way. To meet this need, we designed a concept map and virtual model of a collaborative, virtual cybersecurity escape room. Escape rooms, which are a form of serious games intended to increase knowledge and skills or measure learning outcomes, are scenario-based and interactive in nature. The artifacts described align the serious gaming elements with elements associated with the learning experience to assist game designers and educators in addressing the need for a collaborative virtual space where learners can practice skills associated with cybersecurity in a gamified manner.

Keywords: Cybersecurity · Escape room · Game · Virtual · Education · Training

1 Introduction

This research furthers escape room design, specifically related to cybersecurity in a virtual setting. It describes two artifacts: a concept map and a model of a collaborative cybersecurity virtual escape room. The concept map outlines the relationships of gamification, escape rooms, and learning skills to help future researchers transition content to virtual escape room environments. The map provides a way to represent meanings and ways the meanings are connected [1]. In this particular case, the concept map ties actions in the room to learning skills, enabling designers to adapt future models for a variety of cybersecurity-related learning outcomes.

The prototyped model incorporates the cybersecurity-related skills of social engineering, password security, and binary to create a collaborative virtual experience; therefore, the prototype provides a cybersecurity escape room model that can be used to teach key cybersecurity concepts. The mental map and prototyped model demonstrate how to create a learning experience to support cybersecurity education.

1.1 Motivation

Gamification, which is the use of rules and game design as a motivator to increase learning, has been adopted into a variety of areas, including training and education [2].

K.-K. R. Choo et al. (Eds.): NCS 2021, LNNS 310, pp. 60–73, 2022.
https://doi.org/10.1007/978-3-030-84614-5_6

It revolves around taking a task, such as learning, and making it more attractive to users by structuring it as a game [3]. Building on gamification is the concept of serious games, which are specifically designed to build knowledge, skills, and competencies and have focused learning outcomes. These games can be formatted as board games, strategy games, or action games (games of emergence), or they can require players to follow a set of predetermined actions to complete the game (progression games) [2].

Both games of emergence and games of progression are being adopted to teach cybersecurity-related skills. These games include capture the flag, cyber competitions, and board games [4]. The games have been both physical and virtual and teach topics from binary to phishing [5, 6]. Additionally, game based learning methods have been shown to be well received by students and instructors, as they provide immersive, learner-centered experiences [7].

One of the physical, progression games that is currently in use for cybersecurity education is escape rooms. The approach centers on problem-solving, where students resolve problems posed by the teacher, with the problems framed in the context of a story. The problems are often challenges or riddles, and the learners usually work collaboratively for a specified amount of time [3]. For example, researchers note using a physical escape room as part of their "teach then do" model, where they teach a skill and then have students show their understanding through escape room challenges [8]. This strategy was used to access student understanding of cryptography, data security, wireless protocol manipulation, and embedded systems attacks [8].

1.2 Problem Description

While the escape room game format is popular, the physical escape rooms are challenging to design, difficult to setup, and time consuming to reset. There is also the issue of creating rooms that are sturdy enough to handle multiple uses yet are not too costly to make [9].

Another problem with physical escape rooms is they require players to be in the same physical location. However, learners may be distributed across long distances [10]. One solution is to convert an escape room to a virtual environment. A virtual room would alleviate the need to reset the room, negate the issue of wear and tear on the props, and would allow distributed users to enjoy a collaborative experience.

This brings about the question of how to design a virtual cybersecurity escape room. The purpose of this paper is to first offer a concept map for a virtual escape room design and then utilize this mental model to create a model of a virtual escape room to be used for collaborative cybersecurity training and education. Models for educational simulations, including physical escape rooms, have been documented in the past; however, the proposed artifacts are concerned with maintaining the engagement and problem-solving aspects of physical escape room models while moving the game into a virtual, collaborative setting. The virtual model focuses on cybersecurity content, as the supply of skilled cybersecurity workers is very low [11]. Since one way to build cybersecurity talent is through education, a virtual escape room provides a means to educate learners while building interest in the field [12].

This virtual escape room model provides a way to overcome some of the challenges of physical escape rooms, while still providing learners with a fun educational experience. Hence, we address the following research question:

How to design a virtual escape room to support cybersecurity education?

2 Literature Review

Traditional learning is often teacher centered. This results in less student engagement, as the teacher is the main participant in the learning and serves as the primary diffuser of knowledge [3]. However, one trend is to shift to a learner-centered environment by creating learning by experience, sometimes using tasks and puzzles to promote learning through exploration/experimentation and to meet learner needs [13]. This gamification of learning through serious games helps to create a shift from teacher-centered to learning-centered environments, which also shifts the learners from being extrinsically motivated to being intrinsically motivated [3].

2.1 Benefits of Gamification

Gamification helps facilitate learning, as it can be can adapted to student interests to produce an understanding of content [3]. Part of this is engaging the students in the narrative. Serious games, specifically progression games, such as those found in escape rooms, have storytelling features [2]. This storytelling element is beneficial for learning, as "the advancement in computer technology has helped us in harnessing the story-telling delivered through an interactive and gaming environment" [14].

Interactive environments, which are a part of the gamification of training and education, are also less intimidating than traditional classrooms. The challenges and tests offer the freedom to make mistakes, and the associated rewards, such as badges, are more desirable to students than traditional grades [3]. The format of gamification, which lends itself to flipped classrooms and problem-based learning, is also a benefit. These benefits combine to provide a different motivation for learning than what is seen in traditional classroom rewards. Instead of grades, an extrinsic motivator, students receive a sense of accomplishment, an intrinsic motivator [3].

Indicators of successful gamification of learning include enjoyment, absorption, creative thinking, activation, absence of negative effect, and dominance (level of confidence of the participants) [3].

2.2 Benefits of Physical Escape Rooms

The escape room format is suitable for cybersecurity instruction as it is optimal for cooperative work, e.g. fostering teamwork, creating a high degree of student commitment to meet goals, and increasing engagement [3]. Additionally, the challenges are application based, meaning students progress beyond learning through memorization. This results in deeper learning, as the experimental aspects of the challenges quickly allow students to pinpoint and correct gaps in their knowledge [6].

The format of the serious game also promotes auxiliary skills, such as time management, the ability to prioritize tasks, communication skills, and practice in coping with stress [4]. While the stress produced during play can be negative, the researchers maintain hints and partial solutions, which are an integral part of escape rooms, can offset unhealthy stress and prevent frustration [4].

Escape rooms also increase student motivation and commitment to the task [3]. Researchers contrast this with traditional teaching practices, which they state "fail to attract or motivate students." Instead, escape rooms benefit and enhance academic results.

Other escape room research supports the format's ability to improve academic results. A recent study using pharmacy students found escape room learning was preferred by the majority of students (94.7%) and a majority of students felt they learned better through the format than traditional teaching methods [15].

2.3 Benefits of a Virtual Cybersecurity Learning Environment

Virtual games, specifically cybersecurity-related games, allow for realism and offer a way to study responses [5]. Such games are viewed as immersive, creating a level of engagement that makes it easier to engage, motive, and train users. Researchers note the benefits of a customizable gaming environment, such as can be created using the escape room format. A virtual cybersecurity escape room has the potential to leverage these benefits, helping support cybersecurity education. Specifically regarding cyberse-curity virtual escape rooms, Deeb and Hickey [16] found virtual escape rooms provide authentic learning and are successful in teaching computer security and cryptography to novice students. As far as cybersecurity specific virtual escape rooms, an initial study of a prototype showed overall positive feedback, though participants were only asked to evaluate the playing experience [17]. This research indicates that a virtual escape room model specific to cybersecurity education has the potential to increases student understanding and interest.

2.4 Artifact Requirements

The design research artifacts are intended to support cybersecurity education and virtual escape room design. While Deeb and Hickey provide a 3D escape-the-room game and determine the approach to be successful in providing authentic learning to novice cyber-security students, their research does not offer online collaborative play or document a concept map of the constructs to facilitate further cybersecurity virtual escape room design [16]. Similarly, Löffler et al. focus on transforming an existing physical game into virtual prototype, with little attention on collaborative play and not allowing for distributed play [17].

The goal of the research is to produce a model for a collaborative cybersecurity virtual escape room that retains the positive elements of a physical escape room while incorporating the benefits of a virtual platform. To reach this goal, we use design science research methodology (DSRM), meaning we identified the problem and motivation, defined objectives for a solution, and designed and developed a solution [18]. The design requirements are

- A virtual game that provided an escape room experience.
- A collaborative experience for distributed learners.
- An artifact that supports cybersecurity-related learning objectives.
- An experience that promotes the positive categories of activation (motivation, action, and reward).

- A tool that alleviates a time-consuming reset of the room, negates the issue of wear and tear on the props, and is cheaper than buying and maintaining props.

This research first involves the development of a virtual cybersecurity escape room framework that includes educational and gaming alignments and a primitive prototype of the game.

3   Artifact Design and Development

Our research approach to designing the virtual cybersecurity escape room seeks to satisfy the guidelines for design science in information systems research set forth by Hevner et al. [19]. We began with a survey of the elements of educational game design. In addition to focusing on educational gamification in general, we examined the elements of successful escape rooms and virtual games and the potential problems associated with both. Next, we determined the requirements of the artifact, designed and developed a construct map, used this map to develop a model specific to a collaborative cybersecurity escape room experience, and used virtual prototyping to determine the model's feasibility [18]. This process is based on Peffer et al.'s [18] nominal process sequence, meaning we identified the problem and motivation, defined the objectives of a solutions, designed and developed the artifacts, created a working prototype for demonstration, and defined possible pathways for evaluation and communication.

3.1   Virtual Cybersecurity Escape Room Concept Map

When examining the elements of successful escape rooms and virtual games, we listed relevant constructs and arranged them into a concept map [18]. As Hay et al. [20] note, mapping of relevant constructs and theories is useful in making comparisons and synthesizing information. First, we sought constructs related to activation. Activation is "participation of the learner in the learning process by learning something new" [3].

To build proper activation and promote learning, designs need a motivation, action, and reward. Motivation is defined as the incentive that draws learners into the game. Action is the activities or challenges the user is asked to complete, and the reward is when the learner completes the action and receives the incentive [2]. We added those constructs first to our concept map of processes and structures (see Fig. 1) [21].

Since escape rooms are a form of serious gaming, designers should take into account purpose, content, and play [20]. These elements align with the areas of mentioned by Marín-Vega et al. [2], with motivation mapping to purpose, action mapping to play, and reward mapping to purpose. Content, while not having a direct mapping, has ties to the three. These were added to the map. Eukel and Morrell [13], researching escape room design, describe the learning format as a game loop of overcoming a challenge, finding a solution, and obtaining a reward. Again, there is a relationship among these components (see Fig. 1). We combined the constructs to create a mental model of a virtual cybersecurity escape room.

This simple game loop, along with the additional contributions of Harteveld et al. [22] and Marín-Vega et al. [2] serve as the initial mental model for the virtual cybersecurity

Fig. 1. Virtual cybersecurity escape room concept map.

escape room's design and requirements. Content, which includes the setting, characters, and in the case of serious games, the subject area, impacts the staging of the other elements in the loop. All of the elements and their relationships with each other must be considered in development.

3.2 Design/Refine of a Mental Model

While the concept map provided a scaffold for a virtual escape room, there are other elements to the escape room experience that must be included in the virtual design. Those include characters, a story line, environment, puzzles, hints, rules, distractors (red herrings), participant roles, and time limits.

Also, there are elements needed for a successful virtual gaming experience, including methods to denote achievements and feedback. Examples of these are scoring, progress bars, and timers, which are rewards and indicators of goal attainment [2].

To determine the best path forward, we grouped these elements into categories using the Virtual Cybersecurity Escape Room Concept Map. The goal was to align the items to the elements in the existing model and decide how they relate to a learning experience and a virtual environment. Here, we related content to characters, storyline, environment, distractors, and participant roles; as those elements promote interest and engagement but are not fundamental to the skill being taught. Rather, they contribute to the narrative of the story. For example, to assess a student's understanding of cryptography, the skills could be framed in a narrative centered around either WWII or outer space. These content-related items dictate the framing of the narrative (the setting and characters in the game), but the items do little to impact or assess the learner's understanding.

Puzzles, time limits, and rules are associated with overcoming a challenge. These elements are defined by the skill being taught. The puzzle must assess the knowledge of the skill, while time limits and rules determine the depth of knowledge expected. For example, a beginner may get more time and more lenient rules, but an advanced student might have time drastically reduced or have to adhere to more stringent rules. These items also impact virtual design as these indicators need to be included.

Hints are tied to finding a solution. This aspect assesses the learner's understanding of the task and ability to do the skill. At this point, the game design involves creating a productive struggle without pushing students to the point of frustration [3]. Well-timed hints assist with this.

Finally, obtaining a reward is associated with feedback. Whether a score, progress bar, or a badge, these gauge the learner's understanding of the skill (see Fig. 2). Feedback must be included in a virtual escape room. These skills were applied to the concept map, creating a mental model for future design. This map is a possibility of an outcome and provides a picture of reality to anticipate events [18, 23].

Fig. 2. Virtual cybersecurity escape room concept map with educational and gaming alignments.

As seen above, the educational aspects of the game fell in the outer rings (gaming loop) of the concept map, while the story aspects, which are more flexible and not associated with the specific learning skills, fell into the center (content) section of the frame, inside of the gaming loop. This informs the design as it applies to a virtual setting. Since the center of the mental model is flexible, it can be changed based on the desired theming of the game. However, the outer portions of the mental model should be customized based on skills and student ability, since these areas are tied directly to learning.

This is different than a physical escape room, where designers must consider available resources (props tied to the content/environment) in additional to educational objectives and game goals [13]. The use of a virtual environment lessens the impact of content on game design, as props to create an environment is no longer an issue. This is accounted for in the Virtual Cybersecurity Escape Room Concept Map through larger and darker arrows, representing that a virtual escape room is more skill-driven than narrative-driven. The smaller arrows, pointing away from the content, note that the content can have some impact on game design, but it is less than in a physical environment.

Skill-Driven Elements. As the game is for cybersecurity education, the design is driven by the skills being taught. Therefore, we listed skills that related to cybersecurity. We took into consideration how the skills would fit into a narrative and how easily they could be converted into an online puzzle. Using these as our criteria, we chose pretexting, password security, and binary.

Next, we mapped these skills to puzzles in a virtual space. For pretexting, students find hints in the room that indicated a person's likes or dislikes, which are used for social engineering. For password security, students use the pretexting information they found to answer common password recovery questions. For the last puzzle, they convert a base-10 number to binary, again relating back to a password recovery question. The hints are limited to how to find the needed information in the room and a time limit of 20 min is appropriate for the skills.

For motivation, the game relies on the satisfaction of accomplishing the task and a positive message when students discover the right answer. However, other rewards, such as a digital badge, could be added.

Narrative-Driven Elements. For game design involving a story element, Harteveld et al. [22] outline three steps: 1. Create characters and setting, 2. Create a visual narrative 3. Organize these in the form of dialogue, choice, action. To achieve a virtual escape room, we incorporated these steps in the room's design.

For characters and setting, we needed at least one hero and one villain. We determined the setting should be an apartment in modern times. For, the narrative, we endeavored to answer the questions who, what, when, where, why, and how to weave together a story. The *who* was hero Rush Walker, and the villain organization was the Silent Hand. The *what* was the evidence to help defeat the Silent Hand. The *when* was a 20-min time frame before the participants lost the game. The *where* was an apartment in modern times. The *why* was to save Rush Walker and help find evidence to bring the Silent Hand to justice. The *how* was finding the PIN to unlock the safe deposit box.

Combining the Elements. These elements combine to form the model for the virtual escape room (see Fig. 3). This model demonstrates the relationship of cybersecurity virtual escape room elements, as it shows the relationship among the constructs and how to they relate to cybersecurity virtual escape rooms [24].

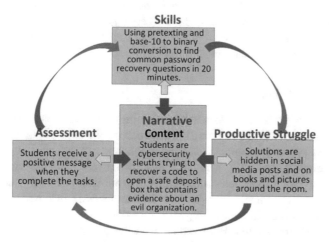

Fig. 3. Virtual cybersecurity escape room model using the concept map.

After populating the mental model with elements for the specific learning model, we created a puzzle chain for the room. This progression chain demonstrates how the narrative-driven elements in the room interact with the skill-driven elements in the room to create a series of puzzle to assess learners' skills (see Fig. 4).

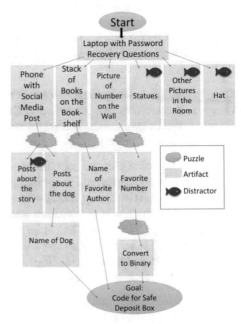

Fig. 4. Puzzle chain.

The puzzle chain is a guide room design in the virtual setting and is an answer key for teachers or others who may be facilitating the play inside the virtual space. The puzzles were arranged in a hybrid fashion, but the design could be adapted to be path-based or sequential, depending on the training needs [9].

3.3   Model Prototyping

To determine if the design could meet the requirements, we prototyped the room in virtual Mozilla Hubs. Mozilla Hubs is a browser-based collaboration platform that allows designers to create and customize virtual environments [25]. This strategy is similar to that of Harteveld et al. [22], who used paper prototypes to create "quick-and-dirty mockups to get quick feedback on design ideas from users without needing to fully implement them." The Mozilla Hubs environment serves as a mockup of a virtual escape room. Link: https://hubs.mozilla.com/jjnf73F/the-silent-hand.

From the existing Mozilla Hubs rooms, we selected one resembling an apartment and began populating the room with items in the puzzle chain. Figure 5 is an example of how the room looks upon entering. The storyboard is in a prominent location to serve as a starting point for play [2].

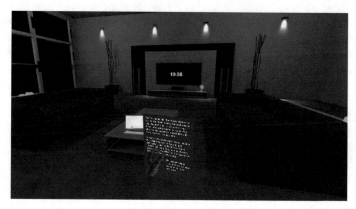

Fig. 5. View upon entering the room.

Due to the room's limited functionality, we added work-arounds to mock-up the interactive aspects of the room. For example, for the users to see the password reset question, we had to embed a link for a Google Form, which opens up a new tab and takes them outside of the Hubs room.

We repeated this strategy for the mobile device that was designed to display a social media post. For it to work, we embedded a link to another site, which we used to create a fake social media post. This post was needed to test social engineering pretexting skills, as the students would have to read the post to find the dog's name for the password recovery question.

The use of these elements could create a problem for learners who have never used this type of virtual environment, as they might have trouble finding and using the open link button for the two props. These elements will be corrected for the game's final design.

Another problem with the mock-up that detracts from the realism of the game is the timer. The only way to embed a timer into the mock-up that would start automatically was to use a GIF. This limits the functionality of the timer, as it only counts down for a few seconds. This is an element that will be corrected in the final game design.Even with the limited functionality of the protype's environment, we were still able to populate the room in an effort to see if the design could meet our requirements.

Aside from work-arounds, the room is indicative of the design. We had no trouble adding the other clues needed for the password recovery questions and hints and rules as specified by our design.

4   Demonstration and Evaluation

Demonstration, which is presenting the artifact in a manner to determine its use for solving one or more aspects of the problem, can involve strategies such as simulations or proof of concept activities [18]. This could be using a workshop to determine the feasibility of an artifact or recruiting stakeholders for expert review, both examples of using demonstrations to determine how well an artifact aligns to design requirements [26, 27].

For the demonstration phase, we will utilize a workshop approach to determine the feasibility of the artifact. A survey will be sent out to 10 high school educators who teach cybersecurity, inviting them to explore the virtual escape room through a link. The survey will consist of questions centering around the initial design requirements. The following questions will be used

- Is the virtual environment similar to an actual escape room?
- Can multiple users enter the space and effectively communicate through audio and chat features; thereby supporting collaborative distributed play?
- Does the design support the skills of pretexting (social engineering), password security, and binary?
- Does the design promote positive categories of activation (motivation, action, and reward)?
- How does the game compare to the physical escape room in regard to alleviating reset of the room, wear and tear on props, and cost of buying and maintaining props?

Open-ended questions will be added to allow participants the opportunity to explain their answers. The responses will be collected using Google Forms, with respondents being asked to rate the degree of their response using a Likert scale ranging from 1 to 5, with 1 being *Does not meet the requirement* and 5 being *Highly meets the requirement*. The scale is intended to assess the degree that respondents feel the artifact aligns with each of the design requirements [28].

Provided that the demonstration indicates the artifact satisfies the design requirements, we will follow Peffer et al.'s [18] DSRM Process Model by having students evaluate the model. This will determine how the game impacts the intended audience.

After comparing various evaluation models, including those by Su et al., Lopez-Belmonte et al. and Hogberg et al., we determined that Su et al.'s framework provides a better coverage of the topics being evaluated and will be used as a guide for developing an evaluation [3, 29]. Additionally, as demonstrated by Couceiro [30], we determined to capture content-related knowledge gained from the activity. This will be achieved using a pretest and posttest.

The demonstration and evaluations will be completed during the next phase of the research.

5   Contribution, Limitations, and Future Work

The use of a virtual platform, as described here for a cybersecurity escape room alleviates a time-consuming reset of a physical room, negates the issue of wear and tear on physical props, and is cheaper than buying and maintaining props.

Aside from making escape room facilitation easier, the research furthers escape room design, specifically in a virtual setting. For example, the concept map informs researchers and game designers on a mental model for a cybersecurity-related, collaborative virtual escape room. It outlines the relationship of gamification, escape room components, and learning skills to help researchers transition content topics to a virtual escape room environment. The concept map also ties actions in the room to learning skills, enabling designers to adapt future models for a variety of educational outcomes.

The research furthers the knowledge of moving the escape room format into a virtual environment by incorporating the skills of social engineering, password security, and binary and making it a collaborative experience; therefore, the cybersecurity content specific prototype provides a virtual cybersecurity escape room model that can be used to teach students social engineering, password security, and binary in a distributed, yet collaborative, format. The game can be utilized in classrooms for training environments to assess learner understanding and promote interest in the field. The prototype demonstrates that such a learning experience is possible to support cybersecurity education.

As the research only allowed for primitive prototyping, future research will involve transitioning the game to an insanitation for further evaluations. Additionally, the future work should evaluate the user and hosts experiences to determine its degree of usefulness in supporting cybersecurity learning outcomes.

References

1. Novak, J.D., Gowin, D.B., Bob, G.D.: Learning How to Learn. Cambridge University Press, Cambridge (1984)
2. Marín-Vega, H., Alor-Hernández, G., Colombo-Mendoza, L.O., Sánchez-Ramírez, C., García-Alcaraz, J.L., Avelar-Sosa, L.: Zeus a tool for generating rule-based serious games with gamification techniques. IET Softw. **14**(2), 88–97 (2020). https://doi.org/10.1049/iet-sen.2019.0028
3. López-Belmonte, J., Segura-Robles, A., Fuentes-Cabrera, A., Parra-González, M.E.: Evaluating activation and absence of negative effect: gamification and escape rooms for learning. Int. J. Environ. Res. Public Health **17**(7), Article no. 7 (2020). https://doi.org/10.3390/ijerph 17072224.
4. Cornel, C.J., Rowe, D.C., Cornel, C.M.: Starships and cybersecurity: teaching security concepts through immersive gaming experiences. In: Proceedings of the 18th Annual Conference on Information Technology Education, New York, NY, USA, September 2017, pp. 27–32 (2017). https://doi.org/10.1145/3125659.3125696
5. Hale, M.L., Gamble, R.F., Gamble, P.: CyberPhishing: a game-based platform for phishing awareness testing. In: 2015 48th Hawaii International Conference on System Sciences, January 2015, pp. 5260–5269 (2015). https://doi.org/10.1109/HICSS.2015.670
6. Ross, R.: Design of an open-source decoder for educational escape rooms. IEEE Access **7**, 145777–145783 (2019). https://doi.org/10.1109/ACCESS.2019.2945289
7. Jin, G., Tu, M., Kim, T.-H., Heffron, J., White, J.: Game based cybersecurity training for high school students. In: Proceedings of the 49th ACM Technical Symposium on Computer Science Education, New York, NY, USA, February 2018, pp. 68–73 (2018). https://doi.org/10.1145/3159450.3159591
8. Streiff, J., Justice, C., Camp, J.: Escaping to Cybersecurity Education: Using Manipulative Challenges to Engage and Educate - ProQuest, October 2019. http://dx.doi.org.elib.uah.edu/10.34190/GBL.19.183
9. Nicholson, S.: The State of Escape: Escape Room Design and Facilities, p. 20
10. Shakeri, H., Singhal, S., Pan, R., Neustaedter, C., Tang, A.: Escaping together: the design and evaluation of a distributed real-life escape room. In: Proceedings of the Annual Symposium on Computer-Human Interaction in Play, New York, NY, USA, October 2017, pp. 115–128 (2017). https://doi.org/10.1145/3116595.3116601

11. Cybersecurity Supply and Demand Heat Map. https://www.cyberseek.org/heatmap.html. Accessed 27 Oct 2019
12. Hoffman, L., Burley, D., Toregas, C.: Holistically building the cybersecurity workforce. IEEE Secur. Priv. **10**(2), 33–39 (2012). https://doi.org/10.1109/MSP.2011.181
13. Eukel, H., Morrell, B.: Ensuring educational escape-room success: the process of designing, piloting, evaluating, redesigning, and re-evaluating educational escape rooms. Simul. Gaming, p. 1046878120953453, August 2020. https://doi.org/10.1177/1046878120953453
14. Pradeep Raj, K.B., Sinha, S., Arvind, R.S., Solanki, D., Lahiri, U.: Design of virtual reality based intelligent storytelling platform with human computer interaction. In: 2018 IEEE/ACIS 17th International Conference on Computer and Information Science (ICIS), June 2018, pp. 142–147 (2018). https://doi.org/10.1109/ICIS.2018.8466457
15. Cotner, S., Smith, K.M., Simpson, L., Burgess, D.S., Cain, J.: 1311. Incorporating an 'Escape Room' game design in infectious diseases instruction. Open Forum Infect. Dis. **5**(suppl_1), S401–S401 (2018). https://doi.org/10.1093/ofid/ofy210.1144
16. Deeb, F.A., Hickey, T.J.: Teaching introductory cryptography using a 3D escape-the-room game. In: 2019 IEEE Frontiers in Education Conference (FIE), October 2019, pp. 1–6 (2019). https://doi.org/10.1109/FIE43999.2019.9028549
17. Löffler, E., Schneider, B., Zanwar, T., Asprion, P.M.: CySecEscape 2.0—a virtual escape room to raise cybersecurity awareness. IJSG **8**(1), Article no. 1 (2021). https://doi.org/10.17083/ijsg.v8i1.413
18. Peffers, K., Tuunanen, T., Rothenberger, M.A., Chatterjee, S.: A design science research methodology for information systems research. J. Manag. Inf. Syst. **24**(3), 45–77, Winter2007/2008 (2007). https://doi.org/10.2753/MIS0742-1222240302
19. Hevner, A.R., March, S.T., Park, J., Ram, S.: Design Science in Information Systems Research, p. 32
20. Hay, L., Cash, P., McKilligan, S.: The future of design cognition analysis. Des. Sci. **6** (2020). https://doi.org/10.1017/dsj.2020.20
21. Storey, M.-A.D., Fracchia, F.D., Müller, H.A.: Cognitive design elements to support the construction of a mental model during software exploration. J. Syst. Softw. **44**(3), 171–185 (1999). https://doi.org/10.1016/S0164-1212(98)10055-9
22. Harteveld, C., Stahl, A., Smith, G., Talgar, C., Sutherland, S.C.: Standing on the shoulders of citizens: exploring gameful collaboration for creating social experiments. In: 2016 49th Hawaii International Conference on System Sciences (HICSS), January 2016, pp. 74–83 (2016). https://doi.org/10.1109/HICSS.2016.18
23. Johnson-Laird, P., Byrne, R.: Mental models: a gentle introduction. Mentalmodelsblog, 02 August 2012. https://mentalmodelsblog.wordpress.com/mental-models-a-gentle-introduction/. Accessed 22 Nov 2020
24. March, S.T., Smith, G.F.: Design and natural science research on information technology. Decis. Support Syst. **15**(4), 251–266 (1995). https://doi.org/10.1016/0167-9236(94)00041-2
25. Welcome to Hubs Hubs by Mozilla. https://hubs.mozilla.com/docs/index.html. Accessed 23 Oct 2020
26. Douma, A.M., van Hillegersberg, J., Schuur, P.C.: Design and evaluation of a simulation game to introduce a Multi-Agent system for barge handling in a seaport. Decis. Support Syst. **53**(3), 465–472 (2012). https://doi.org/10.1016/j.dss.2012.02.013
27. Rusman, E., Ternier, S., Specht, M.: Early second language learning and adult involvement in a real-world context: design and evaluation of the 'elena goes shopping' mobile game. J. Educ. Technol. Soc. **21**(3), 90–103 (2018)
28. Sullivan, G.M., Artino, A.R.: Analyzing and interpreting data from likert-type scales. J. Grad. Med. Educ. **5**(4), 541–542 (2013). https://doi.org/10.4300/JGME-5-4-18

29. Su, C.-H., Chen, K.T.-K., Fan, K.-K.: Rough set theory based fuzzy TOPSIS on serious game design evaluation framework. Math. Probl. Eng. **2013**, e407395 (2013). https://doi.org/10.1155/2013/407395
30. Couceiro, R.M., Papastergiou, M., Kordaki, M., Veloso, A.I.: Design and evaluation of a computer game for the learning of Information and Communication Technologies (ICT) concepts by physical education and sport science students. Educ. Inf. Technol. **18**(3), 531–554 (2013). https://doi.org/10.1007/s10639-011-9179-3

Cyber Security Technology

A Novel Method for the Automatic Generation of JOP Chain Exploits

Bramwell Brizendine$^{(\boxtimes)}$ and Austin Babcock

Dakota State University, Madison, SD 57042, USA
bramwell.brizendine@dsu.edu, austin.babcock@trojans.dsu.edu

Abstract. Jump-Oriented Programming (JOP) is a seldom studied form of advanced code-reuse attacks, very different from return-oriented programming (ROP). JOP identifies snippets of code ending in an indirect jump or indirect call (gadgets), and these are chained together to construct exploits. All applications contain gadgets in executable memory. In this paper we present a mature tool, JOP ROCKET, to facilitate JOP gadget discovery and classification. Additionally, it automates generation a complete JOP chain to bypass Data Execution Prevention (DEP), using a limited virtual machine with emulation. The JOP chain generation utilizes a novel variation to the approach to JOP. Automating JOP chain generation can help provide for automatic detection of vulnerabilities in an application prior to being released, allowing for remediation.

Keywords: Jump-Oriented Programming · Return-oriented programming · Reverse engineering · Software exploitation · Cyber operations

1 Introduction

Vulnerabilities are the bedrock to exploits, and without these, low-level software exploitation is impossible. In response to the ever changing landscape of exploits, we have seen the development of strong and resilient mitigations, such as Data Execution Prevention (DEP), Address Space Layout Randomization (ASLR), Enhanced Mitigation Experience Tool (EMET), and Control Flow Guard (CFG), among others. To thwart mitigations, code-reuse attacks emerged, such as return-to-libc and Return-oriented Programming (ROP). Most modern exploits require code-reuse attacks, to bypass mitigations, e.g. DEP.

Code-reuse attacks use existing regions of code, allowing for code snippets to be chained together, thus achieving arbitrary computation. While ROP is widely known, there is another neglected subset of code-reuse attacks, Jump-Oriented Programming. Although JOP has existed in the literature for a decade, little has been written of it. JOP has been seldom used and poorly understood, to the extent that there were claims it had never before been used in the wild, and that there has never been any publicly available JOP exploit code [1]. JOP was so elusive, it may as well have not existed. Moreover, much practical information on using JOP has never been documented.

K.-K. R. Choo et al. (Eds.): NCS 2021, LNNS 310, pp. 77–92, 2022.
https://doi.org/10.1007/978-3-030-84614-5_7

In this paper we present the Jump-Oriented Programming Reversing Open Cyber Knowledge Expert Tool (JOP ROCKET). Having proposed this framework [2], available on GitHub [3], it has been refined over a period of two years. Going forward, ROCKET will refer to the instantiation of the design science research (DSR) artifact and all its methods. In this paper, we present the artifact itself, which can discover and classify over 150 categories of JOP gadgets, reducing human labor that was tantamount to a finding a needle in a haystack, to an automated task that can be completed in seconds. One ROP researcher wrote that a manual approach to finding just a couple dozen ROP gadgets to meet classification criteria for Turning-complete features took three weeks, having to comb through disassembly from a Solaris libc file [4]. Similarly, absent appropriate tools, doing the same with JOP would likely take longer, owing to the relative paucity of JOP gadgets and its increased complexity.

This paper presents an original method for the automated generation of a JOP chain, using a limited virtual machine that emulates some instructions. The ability to create an automated JOP chain can detect whether software likely would be susceptible to JOP, if a vulnerability were found; these results can be used for remediation.

There are several mature tools that automate discovery of ROP gadgets, such as Mona, ROPgadget, and Ropper [5–7]. Yet for JOP, there has been no comparable tool, prior to JOP ROCKET. Without tooling, a manual process for gadget discovery must be undertaken, requiring multiple tools, as JOP gadgets cannot be found just in the disassembly. Some gadgets are unintended instructions, which often outnumber the intended instructions. A manual approach could take days, yet still not be fruitful.

The problem is the absence of tools to automate and facilitate creating JOP exploits, as the current workflow makes it a manual, tedious, time-consuming process [8–13]. A versatile tool that accounts for the complexity of JOP is required to solve this research problem. Thus, we have proposed, created, and refined a JOP tool, adhering to rigorous DSR guidelines. This paper's contributions include but are not limited to the following: 1) this paper presents a new method for automatic JOP gadget chain generation, resulting in fully developed JOP exploit; 2) this paper presents a JOP gadget discovery tool; 3) this paper expands and refines methods for JOP gadget discovery; 4) this paper presents novel dispatcher gadgets, making JOP much more accessible; 5) this paper presents a system of classification of JOP gadgets.

The remainder of the paper is organized as follows: Sect. 2 provides brief coverage of related work and key concepts; Sect. 3 presents the design and development of JOP ROCKET; Sect. 4 presents evaluation results and further explains contributions of this research; Sect. 5 concludes this paper.

2 Related Work

For decades, there has been an escalating arms race, with both attackers and defenders continuously improving and innovating techniques, alternatively to bypass mitigations or to strengthen defenses. This has necessitated the advent of code-reuse attacks [14]. This section will explore the work both code-reuse attacks and resulting mitigations.

2.1 Code-Reuse Attacks

Code-reuse attacks help bypass mitigations. This first [14] took the form of return-into-libc, which may call a function directly with the needed parameters. That technique expanded to return-oriented programming (ROP) [15–17]. The goal of ROP is to execute borrowed chunks that end in a *ret* instruction; these gadgets exist in the virtual memory of a process. The *ret* instruction pops the address on top of the stack into *eip*. A chain is a series of gadgets in memory. Thus, ROP allows discontinuous instructions to be executed in the order specified by the attacker, subverting W^X.

2.2 Jump-Oriented Programming

Even as of 2015, there were claims [10] there had been no real world JOP attacks. Aside from a handful of articles in the literature, JOP seldom has been written about in detail. All practical information how to do JOP, particularly in a modern Windows environment, has been almost completely absent.

JOP is a state-of-the-art form of code-reuse attacks. The first method is the Bring Your Own Pop Jump (BYOPJ) [15], where a register can be loaded with an address, which is then executed. The next method, the dispatcher gadget paradigm, allows a dispatch table to be crafted in memory [8]. The third approach to JOP [18] is a real-world variation on BYOP, combining functional and dispatcher gadgets as a more labyrinthine chain, allowing for a greater variety of indirect jumps and calls.

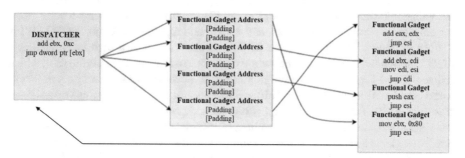

Fig. 1. JOP utilizes a dispatcher gadget and dispatch table to orchestrate control flow.

The dispatcher gadget paradigm [8] is the approach this research favors (Fig. 1). A dispatch table, containing addresses of functional gadgets, is created anywhere in memory. The dispatch table replaces the stack for control flow purposes, providing the order in which the gadgets will occur. Functional gadgets can be viewed as being similar to ROP gadgets, used to deal with mitigations or set up WinAPI calls. The dispatcher is a special gadget that orchestrates control flow. It can advance forwards or backwards in a predictable fashion [8]; it then dereferences and executes functional gadgets. For example, a JOP dispatcher gadget might call a gadget such as *add ebx, 0x4, / jmp dword ptr [ebx]*. If *ebx* contained the address of the dispatch table, the dispatcher gadget would modify *ebx* predictably, advancing 4 bytes. After each functional gadget, the dispatcher is called again, advancing to the next functional gadget until all the functional gadgets have been called, as seen in the diagram.

In the wild, a more limited usage of JOP was to have the register used in the indirect jump, e.g. *eax* from *jmp eax*, and have it point to a *ret*; thus, we could use JOP as if it were ROP under another name [9]. This limited approach to JOP was practical, if an attacker needed an instruction not available from ROP gadgets but available as a JOP gadget. Unfortunately, this approach is impractical for an entire JOP exploit; moreover, it sidesteps one of the JOP goals of avoiding the usage of *ret*, which can be used in some heuristics to detect ROP.

This research focuses on using JOP from the dispatcher gadget method, to develop complete JOP exploits that avoid *ret*. In developing ROCKET and creating JOP exploits, many practical techniques to make JOP work in a modern Windows environment needed to be developed; they were not present in the literature or in the wild.

Similar to ROP, JOP can be used to bypass DEP or ASLR [2]. When bypassing DEP, it is necessary to use the stack, not for control flow purposes, but to place arguments on the stack for Windows API calls, such as VirtualAlloc or VirtualProtect.

2.3 Automatic Generation of ROP Chains

Only a handful of tools perform automatic generation of ROP-chains. The most prominent is Mona [5], which uses determinate recipes to construct ROP chains to bypass DEP. Mona tries to populate registers with appropriate values for VirtualAlloc and VirtualProtect; then it utilizes *pushad* to get all arguments onto the stack, preparing it for the vulnerable function to be called. Ropper [7] is a tool that can create a ROP chain for only VirtualProtect in Windows, and execve and mprotect in Linux. ROPC [19] is a proof of concept Turing-complete ROP compiler with its own high-level language, ROPL, though not intended for real-world exploits. Another effort is Braille [20], which utilizes "Blind Return-Oriented Programming" to develop a ROP chain on a network-based binary with a stack buffer overflow; the requirement is the binary must restart upon crashing. Braille automates a limited BROP attack, generating a ROP chain to transfer the binary over the network, where vulnerability research can then be performed on it. Finally, Roper is a genetic ROP chain compiler [21] that employs genetic programming, in order to automate ROP chain generation. Of the above, only Mona and Ropper are widely used.

2.4 Code-Reuse Mitigations

Enhanced cyber security can be obtained through protections at a binary or system level. These work to stop attacks that utilize memory corruption bugs [22]. These mitigations include DEP, ASLR, EMET, and CFG among others.

Control Flow Integrity

Control flow integrity (CFI) refers to how operating systems implement the natural control flow graph for a program. CFI attempts to determine an application's control flow graph before program execution, allowing the control flow graph to be used to require that control flow adhere to only paths defined in the control flow graph. Thus, if one violated CFI, then this could be detected. CFI can be either fine-grained or coarse-grained. What makes up those paths is defined by different CFI solutions, with varying levels of granularity [23]. In a fully precise static control flow graph, an indirect control flow is permitted only when there is a legitimate trace that follows the edge, avoids malicious attempts at control hijacking, and does not limit functionality. Often real-world implementations rely upon static analysis to create the control flow graph, and the result often is an overly loose, coarse-grained control flow graph [24]. A fine-grained defense is a closer attempt to fully precise control flow graph, but it can be overly restrictive, blocking legitimate paths.

Control Flow Guard and Return Flow Guard
Control Flow Guard (CFG) is Microsoft's coarse-grained CFI implementation. It stores valid addresses in a bitmap and performing a check before every indirect call, to ensure target addresses are valid. CFG can give some defense against ROP when binaries are compiled for CFG. It provides forward-edge CFI, so it can protect against indirect call or jump sites. Microsoft's Return Flow Guard (RFG) provides complimentary support for backward-edge CFI, by providing a software-based shadow stack. With RFG, protection is not assured for return addresses, as valid functions can be called out of context. It is possible to corrupt return addresses on the stack. Researchers [15, 16] have bypassed CFG and RFG. Extended Flow Guard (xFG) is a forthcoming, fine-grained CFI from Microsoft that inserts and checks for compile-time hashes before going to call sites, though in its current implementation, if it fails, it reverts to using CFG [25]. Thus, at present xFG does not provide additional security beyond what CFG already does.

Other CFI Defenses Against ROP
DEP and ASLR are important countermeasures that initially provided some defense against ROP. DEP can be bypassed with minimal effort, assuming there are sufficient ROP gadgets and using tools such as Mona [5]. Other researchers have attempted to create other defenses by detecting ROP, such as with CFI.

Many coarse-grained CFI solutions, such as ROPecker [26], kBouncer [1], ROP-Guard [27], EMET etc. have claimed that they can stop ROP attacks and that Turing-completeness had been eliminated, owing to the reduced codebase. In spite of these claims, Davi and Sadeghi [11, 23] were able to achieve a Turing-complete gadget set on all. Many of these solutions, while they do indeed raise the bar for difficulty, also allow for many more execution paths than are necessary, which a dedicated researcher can take advantage of. Even previously strong countermeasures, such as EMET, which is now natively part of Windows, have been overcome [28]. Some tools provide protection against both ROP and JOP, while some [29] do not utilize heuristics for JOP. This has led control flow integrity being regarded as a potential final solution. Intel's Control-flow Enforcement Tracking (CET) [30] is a hardware-based feature that implements a fine-grained approach to CFI, along with a shadow stack. CET has the potential to be a final solution, but it is not widely deployed, and the majority of existing software is not compiled to support it.

3 Design and Evaluation of JOP ROCKET

This research has resulted in multiple DSR artifacts. These focus on the following: discovery of JOP dispatcher gadgets and functional gadgets, with refinements to the search process; the classification of JOP gadgets; the automatic generation of a complete JOP chain; and the instantiation of the framework itself as a robust tool.

3.1 Design of the JOP ROCKET

DSR was used to guide the development of this tool [31]. ROCKET was implemented in Python, intended for use with PE files. A static analysis tool, ROCKET extracts all executable code from a binary, including all its DLLs, storing them in memory. Each is searched for both dispatcher and functional gadgets. While the binary is being searched, each gadget found simultaneously is classified, with classifications being stored in appropriate data structures. The tool is not proof of concept, but is intended to provide all necessary functionality for real-world JOP exploitation. Thus, the tool is highly versatile and flexible, so a security researcher can customize different options for the search process for both dispatcher gadget and functional gadgets. Because classifications will have been completed after the initial scanning of the binary, users can instantly call upon desired gadgets. Users can then select specific categories or simply select all gadgets found; users may select specific operations and registers affected. The search process can be completed in as little as a minute, although this can increase with larger binaries. Finally, the tool also provides additional functionality to generate a complete JOP chain, producing a Python exploit script. ROCKET presents its features in a simple user interface, requiring minimal keystrokes to complete a search.

3.2 Discovery of Dispatcher Gadgets and Functional Gadgets

ROCKET offers a refinement to the existing algorithm for finding functional gadgets (Fig. 2). First, ROCKET extracts all executable code from a binary, including DLLs; then it searches for opcode patterns for indirect jumps or calls. For instance, it might search for opcodes FF E0 for *jmp eax* or FF D0 for *call eax*. Once found, it uses Capstone disassembly engine to carve out a chunk of disassembly, sending it to function to evaluate fitness and allow for the gadget to be classified. ROCKET will iteratively carve out many chunks of disassembly, each time decrementing the number of bytes used to create the disassembly. Then, regular expressions are utilized to search for specific operations and the registers affected. For instance, it will find all possible JOP gadgets that utilize the *mov* instruction that affect *edi*. Useless gadgets are discarded via filtering. The definition of useless used to assess the fitness of a gadget is the following: 1. There are control flow instructions before the indirect jump. 2. The target operation is of an unusual nature and likely to be impractical, e.g. *xor dword ptr [eax + ebx + 0x43432], bh*. 3. The target operation is redundant, e.g. *mov al, al*.

Dispatcher gadgets are used to predictably advance forwards or backwards in memory, allowing for a dispatch table to be populated accordingly. ROCKET provides an algorithm that looks at results from several classifications of gadgets, such as *add* and *sub*. It then introduces additional criteria to search those results to see if they meet criteria

```
Algorithm 1 JOP Gadget Discovery
Input: Binary File
Output: JOP Gadgets
for reg in Regs:
    for each pos from 0 to textSectionLength do:
        if bytes at pos matches an indirect jmp, then:
            while numOps > 2:
                call disHere(pos, numOps, reg)
                numOps -= 1
Procedure dishere(pos, numOps, reg):
disassembly = generateDisassemblyForChunk()
Reverse potentialGadget
if gadget at disassembly[i] is a valid JOP gadget:
    if gadget at disassembly[i] matches classifier:
        if gadget is not useless:
            saveJOPGadget(pos, numOps, reg, module)
```

Fig. 2. Algorithm for JOP gadget discovery allows for all unintended instructions to be found.

for a dispatcher. Thus, *add edx. 0x4; jmp dword ptr [eax]* would be rejected, as it does not dereference *edx*, while *add edx, 0x8; jmp dword ptr [edx]* would be ideal. ROCKET also checks to see if the advancing operation is done within a short distance from the indirect jump; if it is too far, instructions in intervening lines might make the dispatcher gadget candidate unacceptable.

Discovery of Novel Forms of the Dispatcher Gadget

This research makes novel contributions with gadget discovery, by searching for new variant forms of the dispatcher gadget. First, it searches for dispatchers that use multiplication or shifting left and shifting right by 1 or 2. None of these had previously been considered for dispatcher gadgets, as the values can get very large. However, it is feasible that the attacker could advance forward or backward a very limited number of times. Then they use a functional gadget to subtract or add a large value from the pointer to the dispatch table, thereby resetting the dispatch table. Such a table could accommodate many shifting gadgets. A dispatcher that shifts could be feasible if the attacker gains control over a large expanse of the heap, allowing the dispatch table to have large distances between functional gadgets. While plausible, such a dispatcher gadget would be unlikely to work in most exploits, as finding the right functional gadgets and having control over the heap would be difficult to manage.

ROCKET searches for a hitherto undocumented form of the dispatcher gadget. That is, it searches for an instruction similar to *jmp dword ptr [eax + 0x1]*; this can be generalized as *jmp dword ptr [reg ± offset]*. This form of the dispatcher requires the initial address of the dispatch table to be changed to mirror the offset. The standard dispatcher dereferencing *eax* is made from opcodes FF 20, for *jmp dword ptr [eax]*, while the variant is FF 60 01. Thus, this variant would never be found in the initial search for opcode patterns for traditional dispatchers. With this variant, we search first for the initial two bytes, e.g. FF 60, and then capture what follows next. By searching opcodes, we can discover valuable unintended instructions. However, while promising in theory, in practice there are few dispatchers of this form.

This research presents a novel two-gadget dispatcher, making the requirements for finding a dispatcher less restrictive (Fig. 3). Rather than being reliant upon a scarce gadget, we expand possibilities with two gadgets chained together. The first gadget can modify any register, e.g. *add edi, 0x20; jmp ebp*. The second gadget dereferences the dispatch table, e.g. *jmp dword ptr [ebx]*. ROCKET also provides functionality to discover what we call *empty* jump dereferences; this form of the gadget may exist as only one line, as an unintentional gadget. If expanded to two lines, the instruction would transform into something else. By searching for empty jump dereferences, we nearly always find a jump dereference for all registers. Thus, realistically the only requirement for the two-gadget dispatcher is in satisfying conditions of the first gadget. The two-gadget dispatcher does come at a cost, requiring a third register to be preserved. Two-gadget dispatchers are far more plentiful than their single-gadget counterparts.

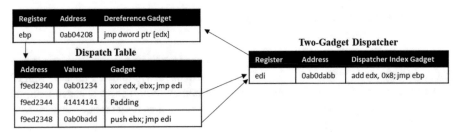

Fig. 3. The novel, two-gadget dispatcher provides needed flexibility for JOP.

The two-gadget dispatcher may dispatch with *call*, unlike the traditional dispatcher. The first gadget of the pair may end in a *call*, e.g. *add ebx, 0x28; call esi*. Because a *call* instruction adds the address of the next instruction to the stack, cleaning up *esp* is necessary. This can be achieved by a *pop reg* right before the dereferencing gadget. Gadgets like *pop ebx; jmp dword ptr [edi]* are plentiful for most registers in any medium or larger size binary. Thus, using a *call* dispatcher is now feasible, although the register used with *pop* would be continuously overwritten.

ROCKET also presents novel forms of a dispatcher that utilize string instructions, such as *lodsd, cmpsd,* and *movsd* to predictably add four to *edi* or *esi*. Used as a single or two-gadget dispatcher, all are effective ways to predictable dispatch.

These novel forms of the dispatcher gadget, none previously documented, can make the previously scarce dispatcher gadget accessible. With our novel dispatchers, it is now possible a find a dispatcher in a medium or large binary. The practical implications are JOP can be extended from academic literature to real-world applications.

3.3 Classification of JOP Gadgets

During the discovery phase, JOP ROCKET makes over 150 distinct classifications based on operation and registers affected, as this classification occurs simultaneously with searching. This research engages in faceted classification for software reuse by organizing knowledge into specific categories [32]. Not only does faceted classification classify the knowledge, in the form of gadgets, but it makes them available for near instantaneous retrieval after classification. This is done with some Turing-complete features, extending it further by adding other granular classifications. Customization for what gadget searches should both include and exclude is possible, allowing search queries to be fine-grained when seeking specific gadgets. Classifications allow users to determine what results are to be saved as text files, with labels denoting functionality. Thus, when a user searches for specific gadgets, they can be found in seconds. Classification is important, as in some larger commercial applications, the number of JOP gadgets could number in the tens of thousands; finding a specific gadget could be time-consuming and tedious. This reduction of human effort allows the security researcher to focus on the exploit, while not missing any gadgets.

3.4 Automatic JOP Chain Generation

While JOP ROCKET is designed to enable a user to easily find all necessary gadgets to manually construct a JOP exploit, this can be labor-intensive work. Our most important contribution is the tool's ability to provide for automatic generation of a complete JOP chain (Fig. 4). ROCKET achieves this using a very limited virtual machine, with emulation of some instructions, using a determinate recipe for chain creation. The JOP chain is saved in Python exploit script; that language is often used in exploit development for payload creation. The JOP exploit allows for DEP to be bypassed. This would then allow for arbitrary execution of shellcode This assumes there is an initial vulnerability to allow the attacker to control EIP and reach the JOP gadgets.

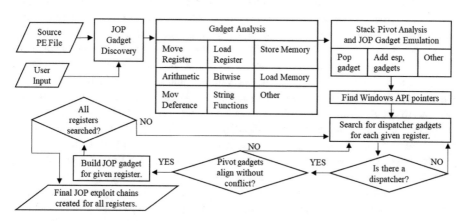

Fig. 4. A flowchart illustrating automatic JOP chain generation.

The generated JOP chain utilizes an alternative approach to JOP that is unique to this research. The chain still utilizes a dispatcher gadget and dispatch table, but in this case all the arguments for a call to VirtualProtect or VirtualAlloc are placed on the stack as part of the payload. The attacker must determine the distance from where the stack arguments land on the stack, and where EIP is located; this distance can be input into ROCKET as a range of acceptable values. Next, a chain of multiple JOP gadgets is created to achieve the desired stack pivot. For instance, if there was a need to pivot 0x1500 bytes to reach the payload, ROCKET would seek a larger stack pivot value, such as 0x950, and then it would make up the difference with smaller stack pivots, to get as close as possible to the desired size without exceeding it. Because there are often smaller stack pivots, it typically reaches the desired stack pivot amount within just 4 bytes; any difference can be made up with padding. A limited virtual machine provides emulation of all instructions that can modify the stack, to determine the precise stack pivot amount that occurs during a gadget as well as a series of multiple gadgets. ROCKET also evaluates the fitness of each gadget for stack pivoting, to ensure there is no conflict with other that must be preserved for the dispatch table or dispatcher. This technique is akin to bouncing from one stack pivot to another, until it reaches a dereferenced jump to a pointer to VirtualProtect or VirtualAlloc, with all arguments on the stack in correct order.

To generate the chain, ROCKET will search for a dispatcher gadget and create a dispatch table with functional gadgets, creating padding to accommodate the distance moved by the dispatcher gadget. For instance, with a dispatcher of *add ebx, 0x8; jmp dword ptr [ebx]*, ROCKET would populate 4 bytes between each address, which itself takes 4 bytes. JOP always uses at least two registers as part of the dispatcher gadget and the functional gadget; ROCKET ensures gadgets selected preserve those registers, preventing them from being clobbered.

JOP could allow for multiple registers being used for control flow purposes, such as a different dispatcher gadget, or a different registers pointing to the dispatcher. Thus, there is not just one possible path, but up to seven different ways. Thus, ROCKET creates all possible chains for registers with sufficient gadgets. However, not all will be useful. For instance, it will generate a JOP chain if there are sufficient gadgets for the stack pivot but no dispatcher gadget. This is because creativity or enlarging the search criteria might yield a valid dispatcher. provided there are gadgets for the stack pivot. Finally, the generated JOP chain utilizes two ROP instructions to load the dispatcher gadget and dispatch table, then transitioning to pure JOP. While JOP can be started with a single JOP gadget, such a starting gadget is not as common, and there two needed ROP gadgets will nearly always be present. Finally, the chain utilizes the stack pivots to then pivot to a call to the API function, with all arguments ready on the stack. Once the API call concludes, it then pivots to the user-supplied shellcode. ROCKET creates a chain to bypass DEP with both VirtualProtect as well as VirtualAlloc and Memcpy.

Nowhere in the literature or the wild is this approach to JOP with multiple stack pivots documented (Fig. 5). One of the features of this approach to JOP is that only a small number of stack pivot gadgets are required, as pivot gadgets can be reused multiple times. This could be as few as one pivot gadget, if no payload size restrictions. Generally, this approach can tremendously simplify the code-reuse attack, if all gadgets are found, including dispatcher, and if there are no restrictions on bad bytes. For a chain to work with minimal adaption, there are a few requirements. First, the null byte must not be a bad character; this is often possible with exploits involving Memcpy. A viable dispatcher gadget must be present, and the binary must support a series of stack pivots with a reachable payload that is predictable. Finally, there must be a pointer in the image executable to VirtualAlloc or VirtualProtect. Other ways of bypassing DEP are possible;

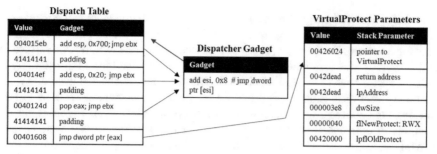

Fig. 5. This novel variation on JOP uses a series of stack pivots to set up and call VirtualProtect, after having moved *esp* a distance of 0x720 bytes.

however, they are not handled by automation. If these conditions are satisfied, then the only adaption required is the initial vulnerability. JOP often will be possible if some of the above requirements are not met, but other problems will need to be solved that are not dealt with by automation. The JOP exploit script provides a strong starting point.

Because JOP is more complex than ROP, automatic JOP chain generation is arguably more challenging. With ROP all instructions end in *ret*, and this is used for control flow, whereas with JOP one must use both a dispatcher gadget and dispatch table. There are several options for each, and each requires registers to be preserved.

3.5 Evaluation Criteria for JOP ROCKET

Evaluation of JOP ROCKET must satisfy several criteria. Firstly, the gadgets it produces must be valid and reliable, meaning all addresses and instructions must be accurate. Next, JOP gadgets must be classified correctly. Next we must demonstrate the generated JOP chain allows for DEP to be bypassed successfully. Finally, this artifact must be evaluated by ensuring it followed design science guidelines [31].

4 Evaluation Results and Contributions

If we satisfy the evaluation criteria and demonstrate the artifact's efficacy, then this research must be considered successful. To show this, we will firstly use several reverse engineering tools to verify the reliability of gadgets, ensuring there are no errors. Next, we will test the artifact in a lab setting, and then we will have others test it, to see if they can use it successfully for its intended purpose, achieving all the above. Finally, we will ensure it meets DSR requirements.

4.1 Validity and Reliability of Results

The classifier for JOP gadgets developed in this work was evaluated rigorously, firstly to see if gadgets correctly aligned with their classifications, and secondly to see if the gadgets produced were accurate. We used several reverse engineering tools, including disassemblers like IDA Pro, Ghidra, and Binary Ninja, and debuggers, such as WinDbg and x64dbg. Multiple tools were used to ensure accuracy, by testing approximately 8,000 artificially generated gadgets to ensure they worked. Inputs containing all possible varieties of valid combinations of gadgets showed all were found and none were missed in a lab environment. Furthermore, over 100, real-world binaries were tested. Verification of disassembly was paramount; addresses and offsets provided for all disassembly for gadgets were shown to be accurate, both for intended and unintended instructions. We verified correct disassembly was produced by going to the addresses in IDA Pro and ensuring the opcodes aligned with what was shown in WinDbg. As expected, the disassembly of unintended instructions would differ in IDA Pro, but they were accurate in WinDbg. Shown below is verification for a gadget containing unintended instruction found by ROCKET. In all cases, the disassembly found for gadgets was correct (Figs. 6 and 7).

```
.text:00495A3D C7 45 98 01 00 00 00        mov    [ebp+var_68], 1
.text:00495A44 8B 4D 8C                    mov    ecx, [ebp+var_74]
.text:00495A47 FF D0                       call   eax
```

Fig. 6. IDA Pro confirms that these are unintended instructions.

```
0:010> u 0x495a42
image00400000+0x95a42:
00495a42 0000                  add    byte ptr [eax],al
00495a44 8b4d8c                mov    ecx,dword ptr [ebp-74h]
00495a47 ffd0                  call   eax
```

Fig. 7. Using the u command in WinDbg, we confirm the unintended instructions are valid.

Reliability encompasses correct classification; we demonstrate all gadgets belong to the correct categories, and none are missed. This was determined by using a manual search process and comparing the results from ROCKET with what was expected. Reliability would also dictate that we do not have multiple instances of the same gadgets repeated, and additionally that there were no unwanted instructions, such as jump conditionals or comparisons in the middle of a gadget; both were demonstrated.

An impactful JOP chain can be defined as one that contains enough required gadgets to be likely to bypass DEP with VirtualProtect or VirtualAlloc. Some chains may lack some elements, such as a dispatcher gadget or pointer to VirtualAlloc, as these sometimes may be found through a manual process. Chains are only ever produced if sufficient, minimum requirements are met. A chain must be adapted then to an initial vulnerability, such as a buffer overflow, which is outside the scope of ROCKET.

Reliability demanded that the generated JOP chain, available as a Python exploit script, worked as intended. To show this, sample binaries were created with certain features, and these were rigorously tested to make sure ROCKET found all unique variations possible. In all, thousands of tests were performed to ensure all aspects of the JOP chain generation worked as intended. Significant testing showed that JOP ROCKET correctly followed a determinate recipe, finding all needed gadgets and then correctly constructing a JOP chain, without missing any gadgets that should have been found. Finally, reliability was assured that the Python exploit script worked without error, and that the generated JOP chain did indeed allow for DEP to be bypassed with only minimal adaptation to include the initial vulnerability.

4.2 Satisfying Design Science Requirements

ROCKET was intended for real-world JOP exploit development, not as a proof of concept prototype; therefore, it must be usable for security researchers and not just work in a controlled experiment. To that end, ROCKET was taught in a doctoral level advanced software exploitation course at Dakota State University; the purpose was to teach students an important code-reuse attack, and ROCKET provides the only pathway to JOP exploitation. Over two years in two separate classes, 32 students used ROCKET to construct unique 64 JOP exploits. The students also created exploits on the same binaries, using ROP. These were challenging exploits that required a bypass of DEP. In one exploit they also needed to find a memory disclosure from a Use-After-Free bug, to develop their own, custom ASLR bypass. Most students were successful with their JOP exploits. Anecdotally, several students said they felt JOP was easier to use than ROP. Strong validation for the efficacy of ROCKET is provided by the fact students used the tool to successfully develop sophisticated JOP exploits.

ROCKET can be evaluated from the lens of Hevner's DSR guidelines. This research yields a viable artifact, in the form of an instantiation, ROCKET itself, encompassing several novel or refined methods. The problem relevance cannot be overstated enough, as without a tool like ROCKET, JOP would be infeasible. For design evaluation, Hevner's guidelines demand utility, quality, and efficacy of the artifact must be shown with rigor and via evaluative methods. This has been demonstrated via single-case mechanism experiment, showing that the gadgets are accurate, that they produce the appropriate offsets, and that the generated JOP chain is effective. With respect to research contributions, the tool itself has provided several novel or refined methods. Different stages of design science research have been iterated through many times over a period of more than two years, incrementally refining the tool.

4.3 Contributions

The paper has satisfied a hitherto unmet need, allowing for an entire class of code-reuse attacks to be possible in a modern Windows environment. Never before was there even so much as a single publicly available example of JOP, and many of the techniques needed to be developed and refined for JOP to actually work. This has culminated in a tool that, as described above, has been used successfully by many individuals, to develop sophisticated exploits that bypass both DEP and ASLR. Second, this paper presented significant refinements to the methods for discovering dispatcher gadgets and functional gadgets. That is, the method of iteratively carving out all possible chunks of disassembly from a desired opcode pattern for an indirect jump or call ensures that all unintended instructions are captured. This research also refines the method for finding a dispatcher gadget, using several novel variants of the dispatcher. With this research, for the first time ever the dispatcher no longer is a scarce. Third, this paper presents a novel method of classification for JOP gadget, with over 150 unique classifications. This level of granularity allows a security researcher to find what they are looking for immediately. In a large application, where there may be thousands of unclassified gadgets to search through, this is a significant reduction of human effort. Forth, this paper present a unique method using the dispatcher gadget and a series of multiple stack pivots, which can allow the security researcher to place arguments for a necessary Windows API, like VirtualProtect, directly on the stack; this can then be pivoted to, immediately executing the API function to bypass DEP and then executing the shellcode. Finally, this paper presents a method for the automated generation of a JOP chain, by using a set of rules and a limited virtual machine with emulation. The result is full JOP exploit in Python to bypass DEP, which requires only minimal modifications, such as introducing the initial vulnerability. This automates the challenging and time-consuming work of creating a JOP chain payload.

4.4 Practical Contributions

Prior to JOP ROCKET and its automatic chain generation, JOP was confined exclusively to the academic literature, almost never done in the wild. In fact, we have published the first real-world JOP exploit [33]. With ROCKET and our research exploring the fundamentals of JOP, we have provided important, novel contributions to make JOP accessible to the security community. The stack pivot method employed in the automatic JOP chain construction requires only a small number of gadgets, many of which can be repeated. For security practitioners performing exploit development, this provides a viable alternative to ROP. Using JOP to weaponize proof of concept exploits forces vendors to remediate buggy software.

5 Final Remarks

This research expands upon what is possible with code-reuse attacks, by making JOP viable in real-world applications, with a mature, well-tested test tool, that has been successfully used to create JOP exploits, both with toy binaries and real-world applications [33]. For students of binary exploitation, using JOP ROCKET provides another way to strengthen cyber skills, as JOP could be used in Capture the Flag (CTF) competitions. Developing zero-day vulnerabilities into proof of concept exploits often requires the use of code-reuse attacks. This work is often carried out by highly skilled security researchers, who make the world's software as secure as possible through bug hunting and vulnerability research. By using the novel techniques developed during our research, security researchers will be able to use ROCKET to determine if software would easily support JOP. If so, they could remediate, making software all the more resilient.

ROCKET is the only framework devoted to JOP, providing advanced features found nowhere else. Its ability to automate the generation of a JOP chain to bypass DEP is a novel contribution that makes an advanced code-reuse attack accessible to all security researchers, allowing JOP to transcend academia, to real-world exploits.

References

1. Pappas, V., Polychronakis, M., Keromytis, A.D.: Transparent ROP exploit mitigation using indirect branch tracing. In: Proceedings of the 22nd USENIX Security Symposium, pp. 447–462 (2013)

2. Brizendine, B., Stroschein, J.: A JOP gadget discovery and analysis tool. S. D. Law Rev. **65**, 540–555 (2020)
3. Brizendine, B.: JOP ROCKET repository. https://github.com/Bw3ll/JOP_ROCKET/
4. Roemer, R.G.: Finding the bad in good code: automated return-oriented programming exploit discovery (2009)
5. Van Eeckhoutte, P.: Corelan Repository for mona.py. https://github.com/corelan/mona
6. Salwan, J.: ROPgadget. https://github.com/JonathanSalwan/ROPgadget
7. Schirra, S.: Ropper. https://github.com/sashs/Ropper
8. Bletsch, T., Jiang, X., Freeh, V.W.: Proceedings of the 6th International Symposium on Information, Computer and Communications Security, ASIACCS 2011 (2011)
9. Checkoway, S., Shacham, H.: Escape from return-oriented programming: return-oriented programming without returns (on the x86). Rep. CS2010–0954, US San Diego, pp. 1–18 (2010)
10. Qiao, R., Zhang, M., Sekar, R.: A principled approach for ROP defense. In: Proceedings of the 31st Annual Computer Security Applications Conference, pp. 101–110 (2015)
11. Davi, L.V.: Code-reuse attacks and defenses. Dissertation (2015)
12. Erdodi, L.: Attacking x86 windows binaries by jump oriented programming. In: Proceedings of the IEEE 17th International Conference on Intelligent Engineering System, INES 2013, pp. 333–338 (2013). https://doi.org/10.1109/INES.2013.6632837
13. Min, J.-W., Jung, S.-M., Lee, D.-Y., Chung, T.-M.: Jump oriented programming on windows platform (on the x86). In: Murgante, B., et al. (eds.) ICCSA 2012. LNCS, vol. 7335, pp. 376–390. Springer, Heidelberg (2012). https://doi.org/10.1007/978-3-642-31137-6_29
14. Shacham, H.: The geometry of innocent flesh on the bone: return-into-libc without function calls (on the x86). In: Proceedings of the ACM Conference on Computer and Communications Security, pp. 552–561 (2007). https://doi.org/10.1145/1315245.1315313
15. Checkoway, S., Davi, L., Dmitrienko, A., Sadeghi, A.R., Shacham, H., Winandy, M.: Return-oriented programming without returns. In: Proceedings of the ACM Conference on Computer and Communications Security, pp. 559–572 (2010). https://doi.org/10.1145/1866307.1866370
16. Roemer, R., Buchanan, E., Shacham, H., Savage, S.: Return-oriented programming: systems, languages, and applications. ACM Trans. Inf. Syst. Secur. **15**, 1–36 (2012)
17. Buchanan, E., Roemer, R., Savage, S., Shacham, H.: Return-oriented programming: exploitation without code injection. Black Hat **8** (2008)
18. M00nbsd: CVE-2020–7460: FreeBSD Kernel Privilege Escalation. https://www.zerodayinitiative.com/blog/2020/9/1/cve-2020-7460-freebsd-kernel-privilege-escalation
19. Pa_kt: A Turing complete ROP compiler (2012). https://github.com/pakt/ropc
20. Bittau, A., Belay, A., Mashtizadeh, A., Mazières, D., Boneh, D.: Hacking blind. In: 2014 IEEE Symposium on Security and Privacy, pp. 227–242 (2014)
21. Fraser, O.L., Zincir-Heywood, N., Heywood, M., Jacobs, J.T.: Return-oriented programme evolution with ROPER: a proof of concept. In: Proceedings of the Genetic and Evolutionary Computation Conference Companion, pp. 1447–1454 (2017)
22. Bania, P.: Security mitigations for return-oriented programming attacks. arXiv Prepr. arXiv1008.4099 (2010)
23. Davi, L., Sadeghi, A.R., Lehmann, D., Monrose, F.: Stitching the gadgets: on the ineffectiveness of coarse-grained control-flow integrity protection. In: Proceedings of the 23rd USENIX Security Symposium, pp. 401–416 (2014)
24. Carlini, N., Barresi, A., Zurich, E., Payer, M., Wagner, D., Gross, T.R.: Control-flow bending: on the effectiveness of control-flow integrity. In: Proceedings of the USENIX Security Symposium (2018)
25. Schenk, M.: eXtended Flow Guard Under the Microscope. https://www.offensive-security.com/offsec/extended-flow-guard/

26. Cheng, Y., Zhou, Z., Miao, Y., Ding, X., Deng, R.H.: ROPecker: a generic and practical approach for defending against ROP attack (2014)
27. Fratrić, I.: ROPGuard: runtime prevention of return-oriented programming attacks (2012)
28. DeMott, J.: Bypassing EMET 4.1. IEEE Secur. Priv. **13**, 66–72 (2015)
29. Chen, P., Xiao, H., Shen, X., Yin, X., Mao, B., Xie, L.: DROP: detecting return-oriented programming malicious code. In: International Conference on Information Systems Security, pp. 163–177 (2009)
30. Intel Corporation: Control-flow Enforcement Technology Preview. https://software.intel.com/sites/default/files/managed/4d/2a/control-flow-enforcement-technology-preview.pdf
31. Hevner, A.R., March, S.T., Park, J., Ram, S.: Design science in information systems research. MIS Q. **28**, 75–105 (2004)
32. Prieto-Diaz, R.: Implementing faceted classification for software reuse. Commun. ACM. **34**, 88–97 (1991)
33. Babcock, A.: IcoFX 2.6 - ".ico" Buffer Overflow SEH + DEP Bypass Using JOP. https://www.exploit-db.com/exploits/49959

Increasing Log Availability in Unmanned Vehicle Systems

Nickolas Carter, Peter Pommer, Duane T. Davis, and Cynthia E. Irvine[✉]

Naval Postgraduate School, Monterey, CA 93943, USA
{nickolas.carter,peter.pommer}@acm.org,
{dtdavi1,irvine}@nps.edu

Abstract. Autonomous multi-vehicle systems are becoming increasingly relevant for many operations. However, logging data within these systems is difficult. In particular, loss of individual vehicles and inherently lossy and noisy communications environments can result in the loss of important mission data. We present a novel distributed ledger protocol, the Unmanned Vehicle System Logging Protocol (UVSLP), that enhances data survival in such systems. We demonstrate the behavioral correctness of this protocol using informal verification methods and tools provided by the Monterey Phoenix project. We further verified the correctness of this protocol through implementation field tests.

Keywords: Cybersecurity · Availability · Distributed ledger · Blockchain · Consensus · Plurality · Monterey Phoenix · UV · UAV · UVSLP

1 Introduction

In distributed systems, the availability of data acquired or generated by distributed nodes can be as critical as cybersecurity requirements for confidentiality and integrity. Consistent and reliable data availability in distributed sensor networks and systems of cooperating autonomous unmanned vehicles (UVs) can be particularly problematic because these mobile systems are often characterized by highly dynamic and potentially unreliable intra- and extra-system communications that impede consolidated mission event log maintenance [1,2]. While operational, the nodes do not communicate with a centralized database system. This paper presents a novel Distributed Ledger Protocol (DLP) approach that fosters mission log survivability of these systems. Called the Unmanned Vehicle System Logging Protocol (UVSLP), our protocol is validated using light-weight formal methods [3]. Experimental results from a real-world implementation on a multi-unmanned aerial vehicle (UAV) system are also presented.

Given the nature of the systems, this protocol differs from traditional distributed database approaches in a number of ways. First, each node maintains a local database in the form of a blockchain that is considered "complete" from the standpoint of that node. The blockchain form fosters immutability of the logged

K.-K. R. Choo et al. (Eds.): NCS 2021, LNNS 310, pp. 93–109, 2022.
https://doi.org/10.1007/978-3-030-84614-5_8

data by associating block hashes to individual blocks and a series of blockchain hashes to the constructed chain. The block hashes ensure immutability of each block while the blockchain hashes ensure immutability of the chain itself. Here, chain *immutability* is with respect to entities external to the system. Unkeyed hash functions were used in this work for demonstration purposes but would be replaced by digital signatures or message authentication codes in practice. A *reconciliation* process is used to bring disparate local blockchains into agreement when divergences are identified. This makes the protocol robust to disconnected networks in which portions of the system temporarily lose contact.

A second difference between the UVSLP and other approaches is that the it relies on local, rather than global, consensus. This enables the protocol to function correctly when system size changes (i.e., new nodes are added or existing nodes depart, fail, or crash). Finally, most blockchain-based DLPs treat the blockchain as additive only [4]. That is, once a block is committed to the chain, it is never removed. The reconciliation process of our proposed DLP systematically removes blocks from divergent chains and then reconstructs a common chain in order to dynamically resolve divergences.

The UVSLP is a novel protocol that has been designed to improve the ability to conduct post-mission forensic analysis of data generated during UAV operations by increasing the likelihood of distributed data surviving the destruction or disconnection of a subset of system nodes. Contributions of this work include:

1. A block-chain-based Distributed Ledger Protocol (DLP) suitable for mission log maintenance for a distributed multi-vehicle autonomous system that suffers from lossy and disconnected communications,
2. Demonstration of a plurality-based approach to consensus within a distributed system,
3. Empirical validation of the approach using light-weight formal methods, and
4. Demonstration of the approach's real-world applicability through simulation and live-flight experimentation on a multi-UAV system.

Section 2 discusses the problem, assumptions, background topics that underpin the protocol, and formal protocol requirements. Section 3 presents a detailed, yet high-level, description of the UVSLP (the full protocol description is provided in [5] and [6]). Section 4 examines the verification process with Monterey Phoenix (MP) [3] to include discussion of abstracted models, MP code, and execution behavior models. Section 4 also includes results of software-in-the-loop (SITL) simulation and live-fly experiments with a full protocol implementation. Finally, Sect. 5 discusses the impacts and implications of this work.

2 Background

This section begins with a description of the target system, then discusses blockchains, consensus, and our protocol requirements, in that order.

2.1   Target System

Our work addresses data logging in autonomous unmanned vehicle systems (UVSs) that incorporate multiple, possibly heterogeneous, vehicles or agents. There is no centralized control, so autonomy is implemented at the vehicle (i.e., agent) level [2]. Each agent independently collects and evaluates sensor observations, maintains its own "situational awareness," and logs selected events and related information. Agents share information only to the extent required to maintain system-wide cohesion.

During typical operation, individual vehicles may fail or crash, groups of vehicles may temporarily detach from the larger system, and agents may be arbitrarily added to or removed from the system. This results in a network with a changing number of nodes that experiences periodic partitioning into sub-networks. Further, even within a connected UVS network or sub-network, communications cannot be considered reliable; point-to-point communications are better understood as a set of dynamic *fair-loss links* between individual agents. That is, any message transmitted by an agent is only probabilistically delivered by the intended recipient agent [7,8].

The UVSLP attempts to overcome logging limitations associated with disjoint and unreliable communications by sharing data in a way that fosters common logs among local UVs. The UVSLP was developed under the following assumptions:

1. The target environment is a distributed system that contains multiple autonomous vehicles.
2. The vehicles operate in a disjoint environment where agents frequently experience network disconnection.
3. The system may experience vehicle failures in which all locally maintained data is lost.
4. The UVSLP ultilizes a cryptographic hash algorithm that effectively precludes hash collisions.
5. The authenticity and confidentiality of all data with which the UVSLP works is ensured by the underlying platform's cryptographic implementation.
6. For now, all agents are trustworthy: Byzantine failures are not considered, and all messages are assumed to be generated and sent in good faith.
7. Each agent creates its own complete blocks, which are comprised entirely of locally-generated information.

2.2   Blockchain

The idea behind blockchains is to use hash functions to cryptographically link consecutive blocks of data [10]. Among other properties, this linkage allows efficient comparison of blockchains and portions of blockchains through comparison of their high-order block hash digests (i.e., if blocks from different blockchains have identical hash digest values, then the blockchains are identical from their origins up to and including those blocks) [11]. The UVSLP leverages this characteristic to identify and reconcile local blockchain divergences between agents.

A blockchain can be characterized as public, permissioned, or private depending on its users and the nature of their interactions [4]. The UVSLP blockchain can be considered private, since it is implemented between trusted agents. It can also be considered to be permissionless since every agent has full write access and participates equally in the commit and reconciliation processes. From the standpoint of the UVSLP, a private, permissionless blockchain does not require proof of work or remuneration (i.e., payment for services), which facilitates its implementation and reduces agent computational workload [12,13].

The UVSLP blockchain further differs from Bitcoin-like blockchains in how the "winning" or "correct" blockchain is selected. With Bitcoin and similar cryptocurrency protocols, the longest blockchain is determined to be correct [9]. In the UVSLP blockchain implementation, the blockchain's length has no bearing. Rather, during the reconcile process the blockchain held by the local plurality is considered the correct blockchain, and divergent local blockchains are reconciled accordingly. In addition, the Bitcoin blockchain places transactions from the losing blockchain back into a data pool to be added to another block for later addition to the blockchain [9], whereas the UVSLP implementation requires that each individual agent create complete blocks and immediately recommits orphaned blocks to the blockchain once the reconcile process is complete.

Perhaps the biggest difference between the UVSLP blockchain and Bitcoin-like blockchains is that the UVSLP blockchain is not additive only. During the reconcile process, locally maintained blockchains are deconstructed from the top down until a common chain is found (i.e., the local chain that maximally agrees with the "correct" blockchain). Following deconstruction, the local blockchain is reconstructed from the common chain to align with the plurality's blockchain.

2.3   Consensus

Among the most important concepts associated with any blockchain protocol is consensus [10]. That is, enough agents must agree on the "correct" blockchain to ensure system-wide consistency. Generally speaking, consensus is formal agreement among a majority of stakeholders with regard to a decision. Unfortunately, full consensus can only be guaranteed under certain conditions [14] that are not satisfied by the target system described in Sect. 2.1. The UVSLP deals with this by relaxing its notion of consensus. Differing locally maintained blockchains are considered "consistent" if a single common chain is obtainable through the reconciliation process [5,6]. This relaxation allows the protocol to gracefully deal with divergences that result from network discontinuities and to achieve full consensus if and when ideal conditions occur. This is particularly important if communications discontinuities result in divergent local blockchains upon recovery. When this occurs, our notion of consistency implies that a post-mission reconciliation process can be used to obtain a single uniform blockchain.

Paxos [15] is a well known family of consensus algorithms for transactional data in which participants are assigned to three roles: proposers, acceptors, and learners. A proposer submits a locally generated value to the acceptors, and if a majority of acceptors agree, the value is committed to the database by the learners. Through this process, it can be demonstrated that any value committed to the database has been agreed upon by the majority [15].

Because the size of our target system is dynamic and communications may be disjoint, the true size of the majority is unknown. Thus, although the Paxos objectives align with those of the UVSLP, it is not directly implementable in our case. Instead, Paxos provides a basis for the UVSLP consensus process. Individual UVSLP agents propose blocks for addition to the blockchain, and, if a local majority (i.e., a majority of agents responding to the proposal) agrees, the block is submitted for addition to the local blockchains [5,6]. Thus, rather than relying on a majority to achieve full consensus, the UVSLP relies on a *plurality* to achieve partial consensus among a UVS subset.

2.4 Protocol Requirements

Per the assumptions described in Sect. 1, the UVSLP is fully implemented on every agent. It consists of a set of handler functions, each triggered by locally or remotely generated events. A correct UVSLP implementation must exhibit the following properties [5,6]:

1. All blocks generated by the protocol will be maintained by at least one agent.
2. No block will exist within the system that was not generated by a participating agent. (This property is partially assured by the agent's underlying cryptographic system for now.)
3. No more than one copy of a particular block will be maintained by any agent at any time. A block can exist within the local blockchain, within a "waiting to be committed" data structure, or within a data structure associated with the reconcile process.
4. If an agent's protocol event handlers are in an idle state, then all blocks maintained by that agent must be present in the local blockchain.
5. In a fully connected system, all blocks will eventually be committed to all locally maintained blockchains.
6. In a fully connected system where all agents have had an opportunity to reconcile blockchains with each other, one uniform blockchain will emerge.

The first four properties relate to the generation and maintenance of blocks and essentially assert that blocks will not be lost, that they will not be erroneously created or duplicated, and that they will all be committed to the local blockchain before the system reaches an idle state. It should also be noted that while the protocol description does not specifically include them, the use of cryptographically sound digital signatures and shared-key message authentication codes can prevent inauthentic blocks from being included and legitimate blocks from being modified or removed. The final two properties relate to the distribution of blocks across the system and assert that, in a fully connected system, all agents will eventually have a copy of every block generated by the system and that the reconciliation process will eventually yield an equivalent blockchain on all agents. Thus, although a uniform blockchain across the entire system at any particular moment cannot be assumed, these last two properties assure that a fully-connected system will eventually obtain a system-wide blockchain among surviving agents (possibly following recovery).

3   Unmanned Vehicle System Logging Protocol (UVSLP)

Depicted graphically in Fig. 1, the UVSLP is designed as a set of event handlers, all of which are implemented on each node in the system. Handlers are triggered by either internally or externally generated events. Data to be recorded in the distributed log is incorporated into blocks that must contain one or more log entries. Blocks are committed to blockchains maintained locally on each node.

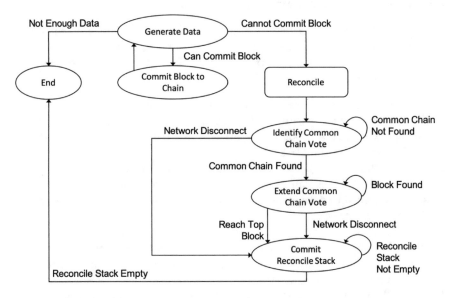

Fig. 1. The Unmanned Vehicle System Logging Protocol (UVSLP).

The protocol is implemented in two main processes: *block generation and commitment*, and *blockchain reconciliation*. These processes are initiated when an agent locally generates data to be entered into the log and adds it to a block under construction. When the block is full, the agent proposes its addition to the blockchain and awaits responses in the form of votes from other agents. If a majority of responses agree to the addition, the block is added to the local blockchain and sent to other agents for addition to their blockchains.

The *blockchain reconcile* process is triggered when a proposed addition is not approved by the local majority or when an agent determines that its local blockchain has diverged. In this process, the reconciling agent removes blocks from its local blockchain until a point of divergence from the local system's common chain is identified. The common chain is then extended to align with that of the local majority, and the removed blocks are readded at that point.

Over the course of an event, one or more nodes may become disconnected from the rest of the system. If they later reconnect with nodes having blockchains that differ, a reconciliation process will eventually be initiated to unify the chains. The reconcile process may occur numerous times throughout a mission. Once a mission is complete, the blocks contained in the locally maintained blockchains can then be analyzed in support of mission analysis and reconstruction.

3.1   Block Generation and Commitment

The *block generation and commitment* process is fully implemented by five event handlers described in detail in [5] and [6]. An abridged description is provided here in the form of two algorithms that describe the process more abstractly.

Generate Block. Described in pseudocode in Algorithm 1, the *generate block* event handler is invoked when an agent generates data to be added to the log. The data is added to the specific block being built by that agent. Once added, the algorithm determines whether or not the block contains enough data to commit it to the blockchain. If the block is large enough, the agent attempts to commit it by invoking the *commit block* event handler. Otherwise, the algorithm enters an idle state so that future log entries can be added to the current block before it is committed.

Algorithm 1. Generate Block

$currentBlock.data \leftarrow currentBlock.data + logEntry$
if $currentBlock$.isFull() **then**
    Invoke **Commit Block** for $currentBlock$
    Create new $currentBlock$
end if

Commit Block. This process is formalized in the pseudocode of Algorithm 2. Once a full block has been generated and submitted for commit, it must be added to the local blockchain and disseminated to other agents in a manner that fosters system-wide consistency. To achieve this the *commit block* component first proposes the addition of the new block to other agents within communications range. The proposal includes a hash digest for the high-order block of the local blockchain to indicate where the proposed block is to be added. All agents that receive the proposal will respond with an "approve" or "disapprove" vote depending on whether or not they approve of the addition. After waiting a prescribed period, the proposing agent tallies all of the received responses. If the "approve" responses outnumber the "disapprove" responses, the agent adds the block to the local blockchain and transmits a commit request message containing

---

Algorithm 2. Commit Block

Clear $proposalVotes$
Create $commitProposalMsg$
$commitProposalMsg.hash \leftarrow localBlockChain.highOrderHash$
$commitProposalMsg.block \leftarrow proposalBlock.ID$
$proposalVoteTimer$.start()
Broadcast($commitProposal$)
while $proposalVoteTimer$.isActive() **do**
    $response \leftarrow$ ReceiveProposalResponse()
    Increment $proposalVotes[response.vote]$
end while
if $proposalVotes[\text{``approve''}] > proposalVotes[\text{``disapprove''}]$ **then**
    Create $commitBlockMsg$
    $commitBlockMsg.block \leftarrow proposalBlock$
    $commitBlockMsg.hash \leftarrow localBlockChain.highOrderhash$
    Broadcast($commitBlockMsg$)
    $localBlockChain$.commit($proposalBlock$)
else
    Invoke **Blockchain Reconcile**
end if

---

the block to the rest of the system. Otherwise, the proposing agent concludes that its local blockchain has diverged from the local majority's common chain, and the *blockchain reconciliation* process is initiated.

Each agent's response to a commit proposal is determined as follows. If the proposal's hash digest matches the high-order block of the receiving agent's local blockchain, then the vote shall be "approve." Otherwise, the vote shall be "disapprove." An "approve" response indicates that the proposing and receiving agents share a common blockchain all the way to the high-order block, and a majority of "approve" responses indicates that the proposing agent shares a common blockchain with a majority of the agents within communications range (i.e., a local majority).

When an agent receives a commit request message, it must determine whether or not the commit is locally valid before adding it to the blockchain. That is, it must confirm that its local blockchain matches that of the proposing agent. Inclusion of the sending agent's high-order local blockchain hash in the request message facilitates this decision. If the receiving agent's local blockchain high-order hash digest does not match the value in the commit request, then the request is ignored. This ensures that externally generated blockchain additions will only build upon a common chain between at least two agents and that a block cannot be added to any local blockchain multiple times.

3.2   Blockchain Reconcile

The *blockchain reconcile* process is initiated to bring an agent's divergent blockchain back into agreement with the local system. The process is split into three distinct phases: *identify common chain*, *extend common chain*, and *commit reconcile stack.* It is fully implemented by 14 event handlers described in detail in [5] and [6]. As with the commit block process, a simplified description of the *blockchain reconcile* process is provided here.

Identify Common Chain. The *blockchain reconcile* process begins with the *identify common chain* phase, as formally illustated in Algorithm 3. A common chain shared with the plurality of local agents is found using a repeated voting process to identify the point of divergence in the local blockchain. In each iteration the reconciling agent broadcasts a query containing its current local blockchain high-order hash value. A receiving agent will respond with "yes" if the hash value is present anywhere in its local blockchain or "no" if it is not. If a majority of responding agents reply with "yes," then the common blockchain has been found, and the *blockchain reconcile* process proceeds to the *extend common chain* phase. If a majority of responses are "no," the top block is removed from the local blockchain and pushed onto a reconcile stack, and the voting process is repeated. If no responses are received, the agent concludes that it has strayed beyond communications range of the rest of the system, and proceeds to the *commit reconcile stack* phase to complete the *blockchain reconcile* process.

Algorithm 3. Identify Common Chain

$commonChainFound \leftarrow False$
$isolatedAgent \leftarrow False$
while not $(commonChainFound$ or $isolatedAgent)$ **do**
    Clear $commonChainVotes$
    Create $containsBlockQueryMsg$
    $containsBlockQueryMsg.hash \leftarrow localBlockChain.highOrderHash$
    $commonChainVoteTimer.start()$
    Broadcast$(containsBlockQueryMsg)$
    while $commonChainVoteTimer$.isActive() **do**
        $response \leftarrow$ ReceiveBlockQueryResponse()
        Increment $commonChainVotes[response.vote]$
    end while
    if $commonChainVotes$.size() $== 0$ **then**
        $isolatedAgent \leftarrow True$
    else if $commonChainVotes["yes"] \geq commonChainVotes["no"]$ **then**
        $commonChainFound \leftarrow True$
    else
        $block \leftarrow localBlockChain.popTopBlock()$
        $reconcileStack$.push$(block)$
    end if
end while
if $isolatedAgent$ **then**
    Invoke **Commit Reconcile Stack**
else if $commonChainFound$ **then**
    Invoke **Extend Common Chain**
end if

Extend Common Chain. In the *extend common chain* phase, the protocol builds upon the common local chain by successively adding blocks that agree with the local plurality's common chain. A voting process is utilized at each iteration to identify the block to be added. The vote begins with a query from the reconciling agent that includes the current local blockchain high-order hash value. Receiving agents search their local blockchains for the query hash and respond in one of two ways if the hash is present. If the query hash is the high-order hash of the local blockchain, the agent responds with a "top block" sentinel value. If the query hash is present but is not the high-order hash value, the agent responds with the block identification for the block immediately following the query hash value in the local blockchain. The query is ignored if the query hash value is not contained in the receiving agent's local blockchain.

The reconciling agent tallies votes at the end of each voting cycle to determine how to proceed. If no responses are received or the plurality indicates that the query hash value is at the top of the common blockchain, the agent determines that the common chain cannot be further extended and proceeds to the *commit reconcile stack* phase. Otherwise, the plurality hash value is chosen for addition to the local blockchain. The agent first checks the reconcile stack for the block, and if it is not present, broadcasts a request for the missing block. Agents receiving the request will reply with a copy of the block, if it is present in their local blockchain. Agents whose local blockchains do not contain the requested block commence blockchain reconciliation processes to rectify their own evident divergence from the plurality blockchain. Once the block is available to the reconciling agent, it is committed to the local blockchain and the next voting cycle is initiated. If no responses to a block request are received, the agent concludes that it has maneuvered beyond communications range, and the *commit reconcile stack* phase is commenced.

Commit Reconcile Stack. The *blockchain reconcile* process finishes with the *commit reconcile stack* phase. In this phase, blocks that were removed from the local blockchain during the *identify common chain* phase and not re-added during the *extend common chain* phase are recommitted. At the commencement of this phase, these blocks are contained in the reconcile stack, a last-in-first-out data structure that ensures that they are recommitted to the blockchain in the same order in which they were originally stored. This phase, depicted in Algorithm 5, pops blocks from the reconcile stack one at a time and commits them to the local blockchain. As each block is committed, a commit message is also broadcast to the rest of the system so that the block can be committed remotely as well. These commit request messages are processed in the same manner as those transmitted during the *block generation and commitment* process.

Algorithm 4. Extend Common Chain

$commonChainComplete \leftarrow False$
while not $commonChainComplete$ **do**
    Clear $nextBlockResponses$
    Create $nextBlockQueryMsg$
    $nextBlockQueryMsg.hash \leftarrow localBlockChain.highOrderHash$
    $nextBlockVoteTimer.start()$
    Broadcast($nextBlockQueryMsg$)
    while $nextBlockVoteTimer.isActive()$ **do**
        $response \leftarrow$ ReceiveNextBlockQueryResponse()
        Increment $nextBlockResponses[response.vote]$
    end while
    if $nextBlockResponses.size() == 0$ **then**
        $commonChainComplete \leftarrow True$
    else
        $pluralityBlock \leftarrow$ SelectPlurality($nextBlockResponses$)
        if $pluralityBlock == TOP\_BLOCK\_SENTINEL$ **then**
            $commonChainComplete \leftarrow True$
        else
            if $reconcileStack.contains(pluralityBlock)$ **then**
                $block \leftarrow reconcileStack.remove(pluralityBlock)$
                $localBlockChain.commit(block)$
            else
                Create $blockRequestMsg$
                $blockRequestMsg.blockID \leftarrow pluralityBlock$
                $blockRequestTimer.start()$
                Broadcast($blockRequestMsg$)
                while $blockRequestTimer.isActive()$ **do**
                    $response \leftarrow$ ReceiveBlockRequestResponse()
                    $blockRequestTimer.active \leftarrow False$
                end while
                if not $response$ **then**
                    $commonChainComplete \leftarrow True$
                else
                    $localBlockChain.commit(response.block)$
                end if
            end if
        end if
    end if
end while
Invoke **Commit Reconcile Stack**

Algorithm 5. Commit Reconcile Stack

---

while not $reconcileStack.isEmpty()$ **do**
    $block \leftarrow reconcileStack.pop()$
    Create $commitBlockMsg$
    $commitBlockMsg.block \leftarrow block$
    $commitBlockMsg.hash \leftarrow localBlockChain.highOrderHash$
    Broadcast($commitBlockMsg$)
    $localBlockChain.commit(block)$
end while

---

4   Verification

Behavior modeling and experiments with an actual multi-vehicle system were used to verify UVSLP correctness. Behavior modeling was conducted with the Monterey Phoenix (MP) platform to test compliance with the required protocol properties. Following that, the protocol was implemented on the Naval Postgraduate School Advanced Robotics and Systems Engineering Laboratory (ARSENL) multi-UAV system. Implementation experiments were conducted in a simulation environment and during live-fly field tests.

4.1   Monterey Phoenix Behavioral Modeling

MP is a lightweight software verification tool that allows the study of complex systems using abstract behavior models [3]. The MP platform was designed to identify logical errors that align with the Small Scope Hypothesis [16]. We utilized MP to evaluate two models to demonstrate the UVSLP's compliance with the six protocol characteristics asserted in Sect. 2.4.

First, we developed a state model and verified that the protocol did not enter an undesired state. Model performance was governed by a set of rules mapped directly to the detailed protocol definition. This model, depicted in Fig. 2, amounted to an abstraction of the protocol that captured the essence of its behavior. Results of exhaustive traces of the model in multiple scenarios using MP confirmed that the protocol did not violate the design requirements [5]. (Each scenario was limited to two loops to prevent combinatoric explosion.)

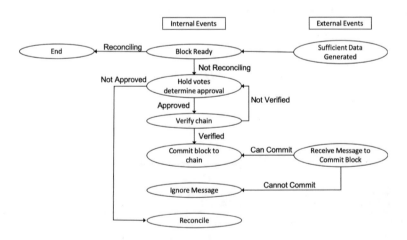

Fig. 2. Abstract protocol state behavior model used in Monterey Phoenix testing.

We also modeled the UVSLP consensus process for MP testing. The resulting abstract consensus model mapped directly to the protocol's consensus process as depicted in Fig. 1. The behavioral model is shown in Fig. 3. We then used MP to run exhaustive traces of the model's performance. (Loop iterations were again limited to two per scenario.) Following this series of tests, trace results again verified that the consensus process performed as desired and that no protocol requirements were violated [5].

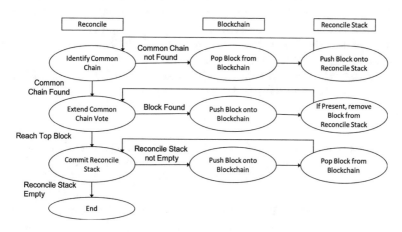

Fig. 3. Abstract consensus process behavior model used in Monterey Phoenix testing.

By examining MP trace results from these models, we were able to verify that the UVSLP behaved as intended. Specifically, these tests provided verification of the correctness of the protocol for small scope examples. Following abstract model testing with MP, a full UVSLP implementation was tested on the ARSENL multi-UAV system.

4.2 Full Implementation Testing

Implementation Overview. The ARSENL system on which the UVSLP was implemented was specifically designed to support experimentation with large numbers of autonomous UAVs (i.e., "swarms") [1]. The system incorporates the fixed-wing and quadrotor unmanned aerial vehicles (UAVs) depicted in Fig. 4. During UVSLP experiments, these vehicles were not controlled by human operators and were subject to communications unreliability as described in Sect. 2.1.

ZephryII fixed-wing UAV [1]   Mosquito hawk quadrotor UAV [17]

Fig. 4. ARSENL multi-UAV system platforms.

The ARSENL system also includes a software-in-the-loop (SITL) simulation capability [18]. This capability allows for full implementation testing in a realistic, physically-based simulation environment. Additional realism is provided by the SITL environment in the form of simulated communications unreliability in which individual packets are probabilistically delivered by each agent.

ARSENL's autonomous capabilities are implemented as a set of Robot Operating System (ROS) nodes running on a companion computer on each vehicle [1]. We implemented the UVSLP within the ARSENL ROS architecture. A single ROS node implements all UVSLP event handlers (i.e., agents). Nodes interacted with the ARSENL infrastructure via ROS message topics. A separate ROS node randomly generated the data to be logged. Log entries consisted of a random integer identification and 16, 32, 64, 128, or 256 bytes of random binary data.

Implementation Testing. SITL simulations with varying levels of packet loss were run to test the UVSLP's performance in lossy-communications environments. Packet loss rates were set to 0, 15, 25, 30, 50, 75, or 100% for these experiments. Note that this testing did not simulate the formation and merging of disjoint networks. (The ARSENL system does not currently provide this capability.) Each test employed five UAVs and ran for 20 min before log entry generation was terminated. All UAVs were then allowed to settle into an idle state (i.e., all commit and reconcile processes were allowed to complete). After each experiment, local blockchains and ROS logs were examined.

Not surprisingly, simulations with lower loss rates yielded more uniform blockchains across the system. Blockchains from simulations with 0, 10, and 15% packet loss converged to a single system-wide common chain. In simulations experiencing between 25 and 30% packet loss, the agents began creating more divergent local blockchains, however significant commonalities remained prevalent. Once packet loss was 50% or more, it became more difficult for a durable common blockchain to emerge, however some commonalities were still noted. As expected, at 100% packet loss each UV maintained a blockchain consisting of only its own blocks. Importantly, all blockchains yielded by these experiments were in compliance with the properties asserted in Sect. 2.4 [6].

Figure 5 (to be read from right to left) displays the final set of blockchains produced by a UVSLP simulation with 30% packet loss. Beginning with a shared "inception block," a significant common chain is present. Divergences occur from this point, but common chains among multiple agents are still present. Further, individual blocks (colored and numbered based on the generating agent and order) are present in multiple chains. Block 2.3, for instance, is present in a common blockchain between Agents 2 and 5 and also in Agent 3's blockchain. These results exhibit exactly the behavior desired of our algorithm.

An interesting feature of the UVSLP is the indirect propagation of blocks across the system. In the experiment depicted in Fig. 5, for instance, we noted blocks 3.5 and 3.6 in the blockchains of both Agent 3 and Agent 5 beyond their common blockchain. (Due to space limitations, not all occurrences are depicted.) Analysis of the ROS logs revealed that these blocks were passed from Agent 3 to Agent 2 and eventually from Agent 2 to Agent 5. Thus, although Agent 3 did not reconcile directly with Agent 5, Agent 3's blocks were indirectly propagated when Agent 2 first reconciled with Agent 3 and later reconciled with Agent 5. Similar block propagation occurred when Blocks 4.4 and 4.5 were passed from Agent 4 to Agent 5 and then from Agent 5 to Agent 2. (These blocks are present in the "hidden" portion of Agents 2 and 5's common chain.)

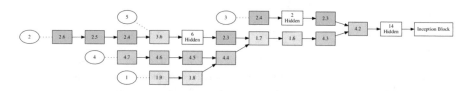

Fig. 5. The final blockchains from a 30% packet loss SITL environment test.

Live-fly testing of the UVSLP implementation was conducted in November 2020 with between five and seven UAVs. Each experiment ran for approximately 30 min while the swarm executed various behaviors. Unlike SITL testing, each vehicle continued to generate log entries until landing, so at least one agent was generating blocks up until the last UAV was on deck. Post-event analysis of local blockchains and ROS logs showed that performance was consistent with that observed in SITL experiments with low to medium packet loss rates [6].

5   Conclusion and Implications

The Unmanned Vehicle System Logging Protocol (UVSLP) enhances data survivability and promotes post-mission data analysis in distributed UVSs by distributing data across all nodes in a system. Data generated during a mission is less likely to be lost, even if the node that generated it perishes. Using Monterey Phoenix (MP), the protocol's behavioral correctness was verified. Field tests validated the functionality of UVSLP and revealed no errors in its design.

In designing the UVSLP, we assumed that the protocol would execute on vehicles subject to possible destruction and network disconnection. As a consequence, the UVSLP can enhance overall system resilience. Although the UVSLP was designed for UVs, its use is not restricted solely to UV systems. In fact, it is suitable for manned vehicles or other distributed systems subject to similar communications constraints. Likewise, although the target domain for UVSLP is airspace, no UVSLP design requirement prevents it from being implemented on land-, space-, or water-based vehicles. Future research will examine mechanisms for the protection of the blockchain, creation of user-friendly management interfaces, scaleability, and disjoint swarm operations.

Acknowledgement. This material was supported in part by the National Science Foundation under Agreement No. 1565443. Any opinions, findings, and conclusions or recommendations expressed are those of the authors and do not necessarily reflect the views of the National Science Foundation.

References

1. Chung, T., Clement, M., Day, M., Jones, K., Davis, D., Jones, M.: Live-fly, large-scale field experimentation for large numbers of fixed-wing UAVs. In: Proceedings of the 2016 IEEE International Conference on Robotics and Automation, Stockholm, Sweden, April 2016 (2016)
2. Coppola, M., McGuire, K., De Wagter, C., de Croon, G.: A survey on swarming with micro air vehicles: fundamental challenges and constraints. Front. Robot. AI **7**, 18 (2020)
3. Auguston, M., Whitcomb, C., Giammarco, K.: A new approach to system and software architecture specification based on behavior models. In: Proceedings of the 3rd International Conference on Model-Based Systems Engineering, Fairfax, VA, September 2010 (2010). https://wiki.nps.edu/download/attachments/604667916/MBSE-2010-final.pdf
4. Nara blockchain white paper. National Archives and Records Administration, February 2019 (2019). https://www.archives.gov/files/records-mgmt/policy/nara-blockchain-whitepaper.pdf
5. Carter, N.: Design and informal verification of a distributed ledger protocol for distributed autonomous systems using Monterey Phoenix. MS thesis, Naval Postgraduate School, Monterey, CA, December 2020 (2020)
6. Pommer, P.: Creation and implementation of a distributed ledger to ensure data survivability across an autonomous system. MS thesis, Naval Postgraduate School, Monterey, CA, March 2021 (2021)
7. Davis, D., Chung, T., Clement, M., Day, M.: Consensus-based data sharing for large-scale aerial swarm coordination in lossy communications environments. In: Proceedings of the 2016 IEEE/RSJ International Conference on Intelligent Robots and Systems, Daejeon, Korea, October 2016 (2016)
8. Bekmezci, I., Sahingoz, O., Temel, S.: Flying ad-hoc networks (FANETs): a survey. Ad Hoc Netw. **11**(3), 1254–1270 (2013)
9. Nakamoto, S.: Bitcoin: a peer-to-peer electronic cash system. Bitcoin, vol. 4 (2008). https://bitcoin.org/bitcoin.pdf
10. Yaga, D., Mell, P., Roby, N., Scarfone, K.: Blockchain technology overview. arXiv preprint arXiv:1906.11078. National Institute of Standards and Technology (2019). https://nvlpubs.nist.gov/nistpubs/ir/2018/NIST.IR.8202.pdf
11. Sherman, A.T., Javani, F., Zhang, H., Golaszewski, E.: On the origins and variations of blockchain technologies. IEEE Secur. Privacy **17**(1), 72–77 (2019)
12. Kannengießer, N., Lins, S., Dehling, T., Sunyaev, A.: Trade-offs between distributed ledger technology characteristics. ACM Comput. Surv. **53**(2), 1–37 (2020)
13. Guegan, D.: Public blockchain versus private blockchain. Documents de travail du Centre d'Economie de la Sorbonne 2017 (2017). https://halshs.archives-ouvertes.fr/halshs-01524440/file/17020.pdf
14. Dolev, D., Dwork, C., Stockmeyer, L.: On the minimal synchronism needed for distributed consensus. J. ACM **34**(1), 77–97 (1987)
15. Lamport, L.: Paxos made simple. ACM Sigact News **34**(4), 18–25 (2001). https://lamport.azurewebsites.net/pubs/paxos-simple.pdf

16. Andoni, A., Daniliuc, D., Khurshid, S., Marinov, D.: Evaluating the "small scope hypothesis". MIT CSAIL (2003). http://citeseerx.ist.psu.edu/viewdoc/download? doi=10.1.1.72.4000&rep=rep1&type=pdf
17. Davis, D., Jones, K., Jones, M., Giles, K.: Advanced swarm UAV capabilities through collaborative field experimentation. Dudley Knox Library, Naval Postgraduate School (2018)
18. Day, M., Clement, M., Russo, J., Davis, D., Chung, T.: Multi-UAV software systems and simulation architecture. In: Proceedings of the 2015 IEEE International Conference on Unmanned Aircraft Systems (ICUAS), Denver, CO, June 2015 (2015)

Testing Detection of K-Ary Code Obfuscated by Metamorphic and Polymorphic Techniques

George T. Harter and Neil C. Rowe[✉]

U.S. Naval Postgraduate School, Monterey, CA 93943, USA
ncrowe@nps.edu

Abstract. K-ary codes are a form of obfuscation used by malware in which the code is distributed across K distinct files. Detecting them is difficult because recognizing the pieces that belong together is hard and provably impossible in general, and the techniques of encryption, metamorphism, and steganography can further obfuscate the code. We built a proof-of-concept K-ary program to test its detectability. It simulated a "keylogger", malware that records keystrokes. We distributed it into parallel obfuscated processes run by a central controller process. We ran both static and dynamic tests to try to detect the keylogger using a variety of parameters. These tests used cosine similarity and clustering methods to correlate pieces of the malware, assuming that using a controller process meant the pieces would have similar code for communications, and that similarity could still be recognized even if obfuscated. Results showed moderate but not perfect success at recognizing our simulated malware. This should provide new tools to detect malware camouflage and evasion.

Keywords: Metamorphic · Polymorphic · Malware · Obfuscation · Evasion · K-ary · Distributed · Malicious · Codes

1 Introduction

Malware (malicious code) is widespread and pervasive. Its goals range from petty crime to espionage and warfare, and some malware attacks have caused billions in damages. For instance, the ILOVEYOU worm has caused over $10 billion in damage [1], and the MyDoom worm is estimated to have caused over $38 billion in damage worldwide [2]. More menacing malware such as the Triton malware, which targets industrial safety-control systems, can endanger human lives [3]. Similarly, the Stuxnet worm sabotaged industrial uranium-enrichment facilities [4].

Anti-malware vendors [5] provide useful defensive tools, but they are inevitably imperfect. An early paper argued that a perfect malware detector cannot be implemented [6], and others have shown a virus can be created for which no automated detector can be implemented without producing false positives [7]. In part this is due to subjectivity in what defines malware; effective malware can be designed to use only components of a benign program. Furthermore, some programs may be considered either malicious or

© The Author(s), under exclusive license to Springer Nature Switzerland AG 2022
K.-K. R. Choo et al. (Eds.): NCS 2021, LNNS 310, pp. 110–123, 2022.
https://doi.org/10.1007/978-3-030-84614-5_9

benign depending on their context. Such ambiguity permits malware authors to make their products more resilient to countermeasures.

Malware can evade detection in many ways. Examples are encryption or steganography to hide their malicious payloads, disguising the malware as legitimate programs, and detecting the presence of analysis programs so as to switch evasion tactics. One method of reducing detectability is to split malware across multiple files, a K-ary design [8]. This aids code obfuscation in that the pieces are less likely to show suspicious features than the whole. The detection of malicious features of K-ary code has been proven to be NP-complete which means it is computationally very difficult [9]. K-ary code can also be enhanced with other stealthy techniques.

Little research has been done on K-ary program designs and few examples have been found outside of laboratory environments. The development of this technique may be like that of ransomware, where early work analyzed asymmetric encryption in a malicious tool [10], but it was not until 2009 that the Archievus ransomware implemented the design in real malware [11]. Today ransomware attacks using asymmetric encryption are common. Similarly, significant K-ary malware might not be seen for many years. If K-ary malware design does prove to be advantageous, defenders should be prepared for its use.

Section 2 gives an overview of malware detection and anti-malware evasion techniques. Section 3 outlines the methods of the experiments, and Sect. 4 describes the results. Section 5 gives concluding remarks and suggestions for future research. More details of our work are in [12].

2 Background

2.1 Malware Detection Methods

Malware detection can be either signature-based or anomaly-based [13]. Signature-based detection is the primary method of anti-malware tools [14]. It recognizes specific artifacts of a file or process which uniquely identify it as malware. The benefits of signature-based detection are efficiency and a low false-positive rate. Its weaknesses are that it cannot detect new malware and it may not recognize malware that has been obfuscated using polymorphic and metamorphic techniques [15, 16]. Executable files with encrypted or compressed bodies offer few signatures because their static code is hidden. The experiments we will describe used novel obfuscated executables, so signature-based techniques for detecting them would have great difficulty.

Anomaly-detection classifies files as malware by recognizing statistical features of known malware [17, 18]. It can use methods such as data mining, machine learning, and statistical analysis to recognize suspicious features. Its main weakness is that it produces more false positives than signature-based methods [19], but these can be reduced by using the consensus of features [13] and a weight-based or rule-based method for combining them [17]. In a weight-based system, indicators of malware are given weights proportional to their strength as indicators, and weighted indicators are summed; if the sum exceeds a defined threshold, the file or process is labeled as malware. A rule-based system uses if-then rules on a file or process [20]; if enough instances of particular patterns found by rules apply, then the file or process is classified as malware. Common clues for anomaly detection are:

- System-call sequences: Application-programmer interface (API) calls used by a program to interact with its host system. These calls may read, create, delete, or modify files on a system, or access a network. Malicious files often exploit these calls for their purposes [18].
- Control flow and call graphs: Although some aspects of a malware may change between instances, most execution flow remains the same [21] and can be recognized [22]. Graph-based detection uses call graphs or flow graphs, and usually parses the executable code after disassembly. It can be quite accurate, but it can be slow and computationally expensive.
- Other clues to malware are code sections marked as writable but not readable, control-flow redirection to a section that does not contain code, gaps between sections of a program, multiple file headers in a single file, and signs of a patched import address table [23].

2.2 Techniques of Malware to Avoid Detection

Metamorphism, Monomorphism, Oligomorphism, and Polymorphism
Metamorphic malware applies code-transformation techniques to the malware body to create versions with identical functionality that are hard to recognize. Metamorphic techniques include [13, 16]:

- Dead-code insertion: Code which does not affect the results is inserted between instructions.
- Independent-code permutation: Instructions are permuted when it does not affect the results.
- Code transposition using added control flow: Segments of code are moved within the file, and control flow is rerouted to jump to them.
- Register renaming: Registers used in instructions are changed consistently.
- Equivalent-code substitution: Blocks of code are replaced by different blocks that do the same thing.
- Subroutine permutation: The locations of subroutines in a file are changed.
- Subroutine inlining: A call to a subroutine is replaced by that subroutine's code.
- Subroutine outlining: Portions of code are made into subroutines and replaced by calls to those subroutines.

Metamorphic malware can be vulnerable to static analysis using graph-based detection and Markov models [22, 24]. It can also be vulnerable to dynamic analysis that examine its behavior when the obfuscation does not change that behavior.

Monomorphic malware usually contains a payload of encrypted malicious code and a built-in decryptor. The decryption key can be changed between instances to make it harder to detect. However, the decryptor itself must remain unencrypted and may be a detectable signature. Furthermore, emulation can see the malware decrypt itself, exposing the malware's payload for analysis.

Oligomorphic malware changes its decryption code between instances or carries code for multiple decryptors [25]. Although oligomorphic is more challenging to detect than monomorphic malware, its decryptors still can be detected.

Polymorphism improves on oligomorphism by using many more decryptors. One study found it in 93% of the malware it examined [26]. It often uses metamorphic code-transformation techniques that permit many versions of the decryption code itself. However, under emulation, polymorphic malware will still decrypt its body, revealing signatures for possible detection, though "on-demand" polymorphism tries to reveal pieces of its code one at a time to reduce signatures [27].

Emulator Fingerprinting and Anti-debugging

Many malware detection and analysis techniques use emulators and virtual machines for dynamic analysis. Malware can possibly recognize them with emulator fingerprinting, and can route execution to non-malicious code or end execution. Fingerprinting clues include [28]:

- Environmental artifacts: User names, files, environment-variable values, and others.
- Timing: Inconsistencies or skews which indicate a process is virtualized.
- Process introspection: Recognition of libraries or hooks that have been injected into the malware code, indicating potential analysis.
- Processor anomalies: Discrepancies in execution that indicate emulation, since it is difficult for an emulator to perfectly imitate a processor or operating system.
- Reverse Turing tests: Lack of signs of human use of a system in such things as the Web-browser history, login times, and the presence of input devices.
- Network artifacts: Discrepancies in what would typically be seen in a network.

Anti-malware tools can use debugging techniques to monitor running processes. On Windows hosts, the functions IsDebuggerPresent and CheckRemoteDebuggerPresent indicate a process being debugged [29], and on Windows NT, flags stored in the NtGlobalFlag variable in the system registry do the same [30]. However, references in the malware code itself to these things themselves are good signatures for anti-malware tools [31].

Anti-disassembly Techniques

Disassembly means reconstructing source code from executable code, and is used by many malware analyzers. It is difficult and even the best disassemblers only correctly reconstruct 40–60% of some programs' executable code when anti-disassembly is used [32]. Malware can exacerbate the difficulty of disassembly by using indirect jumps, exception handling, and other anti-emulation techniques like those mentioned previously. It can also use "flower instructions" that point to sections of valid code at incorrect offsets, but which the malware does not actually execute. Operations can also be obfuscated with exception handling that uses custom signal handlers to redirect execution. These techniques may make disassembly harder, but it still remains possible.

Malware can use steganography (covert channels) to hide code or data [33]. Examples are AdGholas, which embeds malicious JavaScript code in images and other files [34], and ZeroT, which embeds its payload in a set of images [35]. Many anti-malware tools only check specific files like executables for malware, overlooking most steganographic embedding techniques [36].

Packers are file compressors that encode data to reduce its size [37]. Packers can obfuscate since packing eliminates signatures. An estimated 80% of malware uses some form of packer [38]. Because packers often use well-known methods, anti-malware tools and malware analysts often have unpacker programs for them. If no unpacker is available, emulation can wait for the program to unpack itself before analysis.

K-Ary Malware

Another obfuscation technique is dividing malicious code into separate files [39] ("K-ary malware" where K is the number of files) which makes it harder for any subfile to offer a recognizable malware signature [36]. Three methods of execution of such files are:

- Centralized: A single process locates the subfiles, assembles them in its own memory, and executes them [36].
- Sequential actors: The K processes are executed in sequence [9].
- Parallel actors: The K separate processes execute simultaneously.

All three designs can obfuscate against static signature-based detection without needing encryption, steganography, or metamorphism. Covert channels can coordinate processes [39]. A parallel design can also maintain persistence as the malware's multiple processes can restore files or processes that are removed by anti-malware [9]. The most important advantage of all three designs, however, is that it is difficult for analysis to recognize and assemble the pieces of the code even without obfuscation because of all the possible combinations to check.

Nonetheless, signatures are still present with a K-ary design. A centralized design in particular still exposes its full code in memory at times. The sequential and parallel-actor designs require code to coordinate with other processes, and it may be difficult to hide it since a useful signature may be as little as 16 bytes [23]. Statistical analysis may be especially useful for correlating K-ary files or processes. Pieces of executables that have related functionality likely will have related features, as does code written by the same author or using the same tools.

3   Our Methods

Previous work showed that K-ary malware program designs can evade both static and dynamic detection by anti-malware software [36, 39]. Our experiments tested possibly better techniques to detect K-ary malware programs by correlating their pieces. We created an example parallel K-ary program "ManyEyes" to test, simulating a keylogger program that monitors and records the keystrokes made by a user. The program was designed for a Windows operating system. Similar programs have been used by criminal hackers to steal passwords and other sensitive information [40]. To test how traditional obfuscation techniques might be used in K-ary programs, ManyEyes used steganography, encryption, and metamorphism.

We first did static statistical analyses to the program pieces to try to correlate them, using bigram histograms, cosine similarity, and a clustering algorithm. Although the files tested were static, they were processed to simulate being loaded into memory by adjusting their addresses. Experiments were done 1000 times with a variety of malware and benignware combinations. A second set of experiments correlated K-ary memory contents during program execution. Memory dumps of a running instances of ManyEyes were compared statistically to those of other processes.

3.1 Implementation of Our K-Ary Program

Program Design

ManyEyes had subprograms for code gathering, key monitoring, and log aggregation. The code-gathering subprograms took parts of a partitioned executable from secondary storage and executed them in memory. These programs were written in C++ and compiled for x86 architecture systems using the Microsoft C/C++ compiler. In this partly centralized design, a separate code-gathering program was run for each key-monitoring subprogram. The key-monitoring subprograms were Windows executables that monitored key presses and logged them. These were written in C and compiled for a 32-bit x86 architecture using the Microsoft C/C++ compiler. The log aggregators were Python scripts to concatenate log files of around 5 KB.

The code-gathering subprograms were 23 KB. The size of the key-monitoring programs varied because they were generated by a metamorphic engine which increased them from 12 KB to around 45 KB. ManyEyes may be configured with a specified number of key-monitoring subprograms and processing resources. Experiments used a delay of 10 ms between each set of queries, which caused the Windows Task Manager to classify the power use of these programs as "very low".

ManyEyes had five stages: metamorphic code generation, steganographic storage, in-memory execution, keylogging, and log aggregation. Although log aggregation and keylogging may occur simultaneously, each keylogger process hands logs off to an aggregator. The in-memory execution and keylogging used parallel K-ary code design, and the gathering of code portions from steganographic storage emulates the centralized design. Thus ManyEyes uses aspects of all three of the K-ary malware designs, although it is primarily a parallel design.

Obfuscation Techniques

Before executable code pieces were deployed, ManyEyes applied metamorphic transformations to them. For key-monitoring code, we used subroutine reordering and dead-code insertion, and we randomized strings and symbols that could be modified without changing the functionality of the program. These transformations were done at the assembly level before compilation, and the amount of dead code inserted was random.

A design goal of ManyEyes was to prevent unobfuscated key-monitoring code from appearing in storage since anti-malware tools recognize many keylogging methods. To do this, we also used steganography and encryption to obfuscate the code. We hid the code within 24-bit bitmap pictures in the least significant bit of each red, green, and blue value for each pixel in the pictures, and further encoded it with an exclusive-or using a key of 4 to 29 bytes. Each bit of the key was sequentially applied to the least significant bits of each pixel and the key was repeated until exhausted.

The obfuscated key-monitoring programs were 45 kbytes and the entire keylogger payload could be stored in five bitmap files in our experiments. Three of these images were 128 by 209 pixels and approximately 80 kbytes in storage, and the other two were 148 by 197 pixels and about 86 kbytes in storage. The key-monitoring programs used approximately one-ninth of the data in the pictures.

Execution
The key-monitoring subprograms were invoked by the code-gathering subprograms. The latter extracted the former from the picture files, decrypted them, and executed them in memory so they never appeared unobfuscated in secondary storage. This required workarounds in the Windows operating system. We used a method RunPE which spawned a child process in a suspended state, and then wrote code in the Windows Portable Executable format to the child-process memory before releasing its suspension, an implementation was modified from [41]. Cross-process memory writing was done with the Windows applications programming interface with calls to WriteProcess-Memory and VirtualAllocEx. It would have been simpler to concatenate the pieces, write them to a file, and then execute the file from static memory, but that would have revealed the unobfuscated key-monitoring programs in secondary storage.

In the third stage, the key-monitoring programs put into memory were executed to monitor keystrokes and write them to log files. The GetAsyncKeyState function in the Windows applications programming interface was queried repeatedly for each key being monitored. Anti-malware tools could detect this keylogging by looking for frequent calls to the GetAsyncKeyState, but this was obfuscated by splitting the calls over parallel programs with a reduced the number of calls by each process.

In the fourth stage of log aggregation, each aggregator process monitored multiple logs, which it concatenated and wrote to a new location. A hierarchy of log aggregators was used with some working on the aggregations of others, to lower the chances of discovery of their purposes by analyzing their output. To aid this, each log aggregator program also read several random files.

3.2   Implementation of Anomaly-Based Malware Detection

Getting and Preparing Malware Files
Our static-analysis experiments counted bigrams (2-byte sequences) in 1665 benign and 1235 malicious PE executable-file samples. The benign samples were collected from a Windows computer with the permission of the owner. The malicious samples were taken from a corpus of representative files found on real systems [42] by running five anti-malware products (Bit9, Symantec, ClamAV, OpenMalware, and VirusShare) [43] and taking a random sample of the files they identified as malicious. The benign samples had 32,4252,414 bigrams (two-byte sequences) and the malicious samples had 139,696,687

bigrams. Both sets of data contained every possible bigram of the 65,536 possible. The average benignware file was 407 KB and the average malicious file was 218 KB.

To test our detectability of our split code by anomaly-detection methods, 1,235 malware samples were used and their executable code was split into five parts for the first 1,000 runs of this experiment, and into fifteen parts for the second 1,000 runs of this experiment. In each experiment run, a single sample of the malware was selected. We also selected some benignware samples from the same corpus.

Most modern operating systems use the security technique of address-space layout randomization, which causes memory references and virtual-memory placements to differ between instances of a program. To simulate it, we added a random integer to memory references marked for modification in the relocation table for each malware part. For consistency, the benign programs in our experiments also used this simulated memory mapping.

Clustering of N-Grams

Our static-analysis experiments correlated our K-ary code pieces by clustering them based on their similarity. The metric was the cosine similarity (normalized inner product) of their histograms of byte sequences of length N (N-grams). N-gram statistical analyses are often used in deep packet inspection for cybersecurity. We clustered all the malware and benignware pieces and measured how often malware pieces were placed in the same cluster. Clustering was done by HDBSCAN, a hierarchical density-based clustering algorithm based on DBSCAN [44]. HDBSCAN can discover clusters of varying densities and requires less prior knowledge about the nature of the clusters than related algorithms.

For each of 1,000 runs of each experiment, a malware program and 200 benign programs were randomly selected. This simulated the proportion of malware to benignware that is likely to run simultaneously. For each run, the histograms for each of the K pieces of the malware being tested and the 200 benign samples were paired and their cosine similarity was calculated. The benignware were not split, as we expect they would not be split on a real system.

Success of K-ary malware partitioning was measured by the number of malware pieces in the cluster with the most malware pieces. True positives were malware pieces in that cluster, false positives were benignware in that cluster, and false negatives were malware pieces in another cluster. Using these measures, we calculated accuracy, precision, recall, and the F-score for the identification of malware pieces in each experiment. Accuracy measures how many benign programs are not placed into clusters with malware parts and how many malware parts are. Recall measures how many malware parts are in clusters with other malware parts. Precision measures how many few benign programs are clustered with malware parts. The F-score is the geometric mean of recall and precision.

Correlating Live Processes

A second set of dynamic-analysis experiments correlated the running processes of ManyEyes on a Windows system. In these we captured 120 memory dumps of benign processes and two of ManyEyes. The dumps captured were only of memory regions marked as executable. One ManyEyes version was composed of five code-gathering subprograms, five key-monitoring programs, and a single log aggregator. Since this configuration resulted in six parallel processes, we call it the 6-piece ManyEyes instance. The second version was composed of fifteen code-gathering subprograms, fifteen key-monitoring subprograms, and five log-aggregators. Because this resulted in twenty parallel processes, we call it the twenty-piece ManyEyes instance. Experiments were repeated without entropy analysis, without metamorphism, and without both.

Collecting data from live processes was challenging. Data can represent uninitialized portions of the memory with zeros or data from previous processes. Some of it can be ignored in analysis by examining its entropy and excluding sequences with very low or very high entropy. We examined sixteen-byte chunks of data and removed portions that had entropies higher than 0.7 or lower than 0.3 of the maximum value. To produce the memory dumps, we created a tool named "MemoryScanner" to capture executable regions of memory on a Windows system. We chose to build our own tool to get greater control of what data it accessed. The tool uses the Windows ReadProcessMemory applications programing interface and avoids memory regions protected by PAGE_NOACCESS and PAGE_GUARD permissions. The tool does not address potential image "smear" from reading memory while it is changing.

4   Results

Table 1 shows the average results of our static analysis experiments with malware split into five or fifteen parts. High values of the statistics of cluster accuracy, precision, recall, and the F-score indicate clusters had more malware parts. We see that about half of malware pieces when five parts were used were put in a single cluster, which indicates that clues to malware occur even after splitting and obfuscation. For splitting into fifteen pieces, the precision increased significantly, but the recall decreased. This suggests that both positive and negative clues decrease as sizes become smaller.

Table 1. Statistics for the split-malware experiments.

Number of parts into which the malware was split	Average accuracy	Average precision	Average recall	Average F-score
5	0.9827	0.6927	0.5192	0.5935
15	0.9488	0.7935	0.3593	0.4946

Figure 1 compares accuracy, precision, recall, and F-score for memory dumps of processes using a 20-piece instance of ManyEyes malware with fifteen key-monitoring processes and five log aggregators. When metamorphism was enabled and extreme-entropy code was not considered, the recall was much higher than in the other runs. This is surprising because the metamorphic engine tries to make correlation harder, and may be due to the limited variety of its dead-code instructions. When metamorphism was disabled so identical code occurred in the parallel key-monitoring subprograms, the recall increased when extreme entropy was removed.

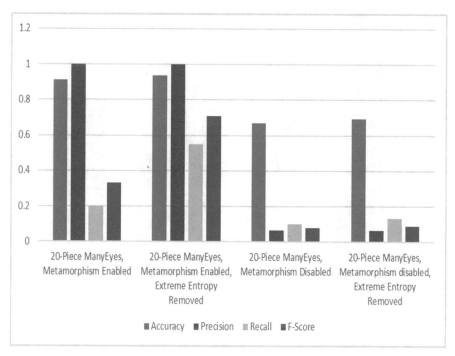

Fig. 1. Performance of clustering of 20-piece ManyEyes processes

Figure 2 shows corresponding statistics for 6-piece ManyEyes experiments. Precision was much higher than in the 20-piece experiments when metamorphism was disabled. The F-score was highest when metamorphism was disabled and extreme entropy was not removed. The recall was highest when metamorphism was disabled regardless of whether extreme entropy was removed. These tests gave expected results when disabling metamorphism and removing extreme entropy, resulting in better correlation of ManyEyes' processes. The precision of each experiment was perfect except when metamorphism was disabled and extreme entropy was removed, which was the worst-performing of the 6-piece tests. Unlike with the 20-piece tests, removing extreme entropy did not improve performance. Uninitialized memory can contain code from other processes, which means that entropy analysis alone may not suffice to isolate a live process' executable code. These tests show that the correlating K-ary processes using statistics is possible but not guaranteed.

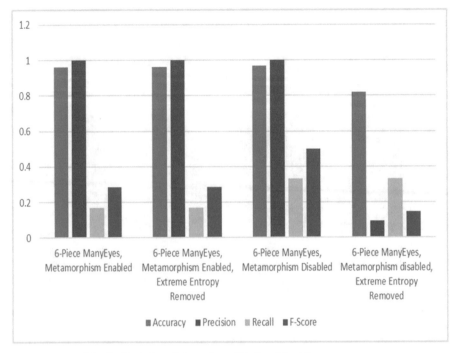

Fig. 2. Statistics of clustering of 6-piece ManyEyes processes

5 Conclusions and Future Work

This work explored the potential stealth of K-ary malicious code combined with traditional code-obfuscation techniques. In our implementation we saw that steganography, encryption, and metamorphism can increase K-ary malware stealth. Our results showed that they can protect malware by reducing its signatures and preventing correlation of its pieces when partitioned, but that such concealment was not perfect.

This work used a statistical method for correlating pieces of executables using cosine-similarity comparison of bigram histograms on running processes. Our experiments showed that it was an effective metric. Memory dumps of ManyEyes during execution were also compared to memory dumps of other benign programs and showed some ability to correlate K-ary malware processes, but the correlation was lower than with the static images.

One possible direction for K-ary malware design is for parts of the malware to monitor the others and restore any pieces that are removed by a system administrator or anti-malware program [9]. Shifting or permuting code across files could also protect K-ary files against signature-based detection. Grammars can automatically generate K-ary codes [45]. Techniques that would allow K-ary code to collaborate stealthily are another possibility.

The degree to which K-ary code will be used by malware authors to obfuscate their malware remains to be seen. Our results suggest that statistical analysis can however decrease the threat of K-ary code.

References

1. Ingram, M.: "Love-bug" virus damage estimated at $10 billion. World Socialist website (2000). https://www.wsws.org/en/articles/2000/05/bug-m10.html
2. Palmer, D.: MyDoom: the 15-year-old malware that's still being used in phishing attacks in 2019. ZDNet (2019). https://www.zdnet.com/article/mydoom-the-15-year-old-malware-thats-still-being-used-in-phishing-attacks-in-2019. Accessed 26 June 2019
3. Giles, M.: Triton is the world's most murderous malware, and it's spreading. Technology Review (2019). https://www.technologyreview.com/2019/03/05/103328/cybersecurity-critical-infrastructure-triton-malware. Accessed 9 Mar 2019
4. Zeter, K.: An unprecedented look at Stuxnet, the world's first digital weapon. Wired (2014). https://www.wired.com/2014/11/countdown-to-zero-day-stuxnet. Accessed Nov 2014
5. Liu, S.: Security software – statistics & facts (2020). https://www.statista.com/topics/2208/security-software. Accessed 9 Jan 2020
6. Cohen, F.: Computer viruses: theory and experiments. Comput. Secur. **6**(1), 22–35 (1987)
7. Chess, D., White, S.: An undetectable computer virus. In: Proceedings of the Virus Bulletin Conference, vol. 5, pp. 1–4 (2000)
8. Moubarak, J., Chamoun, M., Filiol, E.: Developing a K-ary malware using blockchain. In: Proceedings of NOMS 2018–2018 IEEE/IFIP Network Operations and Management Symposium, pp. 1–4 (2018)
9. Filiol, E.: Formalisation and implementation aspects of k-ary (malicious) codes. J. Comput. Virol. **3**(2), 75–86 (2007)
10. Young, A., Yung, M.: Cryptovirology: extortion-based security threats and countermeasures. In: Proceedings of the 1996 IEEE Symposium on Security and Privacy, pp. 129–140 (1996)
11. O'Kane, P., Sezer, S., Carlin, D.: Evolution of ransomware. IET Netw. **7**(5), 321–327 (2018)
12. Harter, G.: Metamorphic and polymorphic techniques for obfuscation of k-ary malicious codes. M.S. thesis, U.S. Naval Postgraduate School (2020). http://calhoun.nps.edu
13. Desai, P., Stamp, M.: A highly metamorphic virus generator. Int. J. Multimed. Intell. Secur. **1**(4), 402–427 (2010)
14. Preda, M.D., Di Giusto, C.: Hunting distributed malware with the κ-calculus. In: Owe, O., Steffen, M., Telle, J.A. (eds.) Fundamentals of Computation Theory . LNCS, pp. 102–113. Springer, Heidelberg (2011). https://doi.org/10.1007/978-3-642-22953-4_9
15. Saeed, I., Selamat, A., Abuagoub, A.: A survey on malware and malware detection systems. Int. J. Comput. Appl. **67**(16), 25–31 (2013)
16. You, I., Yim, K.: Malware obfuscation techniques: a brief survey. In: Proceedings of the 2010 IEEE International Conference on Broadband, Wireless Computing, Communication, and Applications, pp. 297–300 (2010)
17. Gaikwad, P., Motwani, D., Shinde, V.: Survey on malware detection techniques. Int. J. Mod. Trends Eng. Res. **21**(7), 1–25 (2015)
18. Bazrafshan, Z., Hashemi, H., Fard, S.M.H., Hamzeh, A.: A survey on heuristic malware detection techniques. In: Proceedings of the 5th IEEE Conference on Information and Knowledge Technology, pp. 113–120 (2013)
19. Wong, W., Stamp, M.: Hunting for metamorphic engines. J. Comput. Virol. **2**(3), 211–229 (2006)
20. Schmall, M.: Heuristic techniques in AV solutions: an overview (2002). https://www.symantec.com/connect/articles/heuristic-techniques-av-solutions-overview
21. Bachaalany, E., Koret, J.: The Antivirus Hacker's Handbook. Wiley, New York (2015)
22. Eskandari, M., Hashemi, S.: Metamorphic malware detection using control flow graph mining. Int. J. Comput. Sci. Netw. Secur. **11**(12), 1–6 (2011)
23. Szor, P.: The Art of Computer Virus Research and Defense. Addison-Wesley , Reading (2005)

24. Lee, J., Jeong, K., Lee, H.: Detecting metamorphic malwares using code graphs. In: Proceedings of the 2010 ACM Symposium on Applied Computing, pp. 1970–1977 (2010)
25. Schiffman, M.: A brief history of malware obfuscation: part 1 of 2. Cisco (2010). https://blogs.cisco.com/security/a_brief_history_of_malware_obfuscation_part_1_of_2
26. Webroot: Webroot threat report (2019). https://www-cdn.webroot.com/9315/5113/6179/2019_Webroot_Threat_Report_US_Online.pdf
27. Alvarez, R.: Dissecting a metamorphic file-infecting ransomware (2018). https://www.youtube.com/watch?v=vJ08_6CCd6g. Accessed 26 Mar 2018
28. Bulazel, A., Yener, B.: A survey on automated dynamic malware analysis evasion and counter-evasion: PC, mobile, and web. In: Proceedings of the 1st ACM Reversing and Offensive-oriented Trends Symposium, p. 2 (2017)
29. Gao, Y., Lu, Z., Luo, Y.: Survey on malware anti-analysis. In: Fifth IEEE International Conference on Intelligent Control and Information Processing, pp. 270–275 (2014)
30. Kulchytskyy, O., Kukoba, A.: Anti debugging protection techniques with examples (2019). https://www.apriorit.com/dev-blog/367-anti-reverse-engineering-protection-techniques-to-use-before-releasing-software. Accessed 23 May 2019
31. Chen, W.: Encapsulating antivirus (AV) evasion techniques in the Metasploit framework (2018). https://www.rapid7.com/globalassets/_pdfs/whitepaperguide/rapid7-whitepaper-metasploit-framework-encapsulating-av-techniques.pdf. Accessed 9 Oct 2018
32. Popov, I., Debray, S., Andrews, G.: Binary obfuscation using signals. In: Proceedings of the USENIX Security Symposium, pp. 275–290 (2007)
33. Rowe, N.C., Rrushi, J.: Introduction to Cyberdeception. Springer, Cham (2016). https://doi.org/10.1007/978-3-319-41187-3
34. SentinelOne: Hiding code inside images: how malware uses steganography (2019). https://www.sentinelone.com/blog/hiding-code-inside-images-malware-steganography. Accessed 4 July 2019
35. Yoon, S.: Steganography in the modern attack landscape (2019). https://www.carbonblack.com/2019/04/09/steganography-in-the-modern-attack-landscape. Accessed 9 Apr 2019
36. Ramilli, M., Bishop, M.: Multi-stage delivery of malware. In: Proceedings of the 5th IEEE International Conference on Malicious and Unwanted Software, pp. 91–97 (2010)
37. Arntz, P.: Explained: packer, cryptor, and protector (2017). https://blog.malwarebytes.com/cybercrime/malware/2017/03/explained-packer-crypter-and-protector
38. Guo, F., Ferrie, P., Chiueh, T.: A study of the packer problem and its solutions. In: Lippmann, R., Kirda, E., Trachtenberg, A. (eds.) RAID 2008. LNCS, vol. 5230, pp. 98–115. Springer, Heidelberg (2008). https://doi.org/10.1007/978-3-540-87403-4_6
39. Ramilli, M., Bishop, M., Sun, S.: Multiprocess malware. In: Proceedings of the 6th IEEE International Conference on Malicious and Unwanted Software, pp. 8–13 (2011)
40. Solairaj, A., Prabanand, S., Mathalairaj, J., Prathap, C., Vignesh, L.: Keylogger software detection techniques. In: Proceedings of the 10th IEEE International Conference on Intelligent Systems and Control (ISCO), pp. 1–6 (2016)
41. Zer0Mem0ry: RunPE (2016). https://github.com/Zer0Mem0ry/RunPE/blob/master/RunPE.cpp
42. Garfinkel, S., Farrell, P., Roussev, V., Dinolt, G.: Bringing science to digital forensics with standardized forensic corpora. Digit. Investig. **6**, S2–S11 (2009)
43. Rowe, N.: Finding contextual clues to malware using a large corpus. In: Proceedings of the ISCC-SFCS Third International Workshop on Security and Forensics in Communications Systems, Larnaca, Cyprus (2015)

44. McInnes, L., Healy, J.: Accelerated hierarchical density clustering. In: Proceedings of Workshop of the IEEE International Conference on Data Mining, pp. 33–42 (2017)
45. Gueguen, G.: Van Wijngaarden grammars and metamorphism. In: Proceedings of the Sixth International Conference on Availability, Reliability, and Security, Vienna, AT, pp. 466–472 (2011)

Enhancing Secure Coding Assistant System with Design by Contract and Programming Logic

Wenhui Liang, Cui Zhang, and Jun Dai[✉]

Department of Computer Science, California State University, Sacramento 6000 J Street, Sacramento, CA 95819, USA
{wenhuiliang,zhangc,jun.dai}@csus.edu

Abstract. The system titled Secure Coding Assistant was developed to automate early detection for a subset of the Java secure coding rules specified by the SEI CERT at the Carnegie Mellon University. This system can help Java programmers significantly reduce security vulnerabilities in their code caused by the violations of secure coding rules. Since other software defects can also lead to security vulnerabilities, efforts have been taken to extend Secure Coding Assistant aiming at empowering programmers to detect, locate and remove code errors during coding time. This paper presents an enhancement to Secure Coding Assistant by a combination of Design by Contract and Programming Logic. Java programmers using this system are advised to provide their design contracts, i.e., logic assertions, for program structures of methods, if-then-else statements and while-loop statements. The design contracts defined by programmers can be automatically checked at the time of their program execution. To further facilitate the process of detecting and locating of code errors, using the programmers-defined design contracts, sub-design contracts can be automatically generated by the system based on the inference rules for the if-then-else statement and the while-loop statement in programming logic. The sub-design contracts generated by the system can also be automatically checked at dynamic time. In addition, based on the assignment axiom and the inference rule for the sequence statement in programming logic, the weakest pre-conditions of certain assignment sequences can be automatically generated from the post-conditions of the sequences, enabling programmers to statically analyze the correctness of the corresponding design contracts they specify. With the enhancement presented, Secure Coding Assistant can assist programmers for the early detections of not only secure coding rule violations but also errors in code. These early detections are performed in unison with the coding process to pursue software security.

Keywords: Secure coding · Software security · Design by contract · Programming logic

1 Introduction

Network attacks have become more and more common in recent decades. Cybersecurity Ventures predicts that in the next five years, the cost of global cybercrime will grow at

K.-K. R. Choo et al. (Eds.): NCS 2021, LNNS 310, pp. 124–140, 2022.
https://doi.org/10.1007/978-3-030-84614-5_10

an annual rate of 15%, reaching \$10.5 trillion USD by 2025, more than \$3 trillion USD in 2015 [4]. Since 2018, McAfee estimate the cost of cybercrime worldwide to be more than \$1 trillion. They estimate that the money loss caused by cybercrime is about 945 billion US dollars. In addition, global cybersecurity spending is expected to exceed \$145 billion by 2020. Today, it is a trillion dollar drag on the global economy [18]. Software enterprises face many challenges in diverse areas such as system security, application security, sensitive information protection. Developing software security is an inevitable trend.

To address the increasing needs for software security, the SEI CERT at the Carnegie Mellon University specified secure coding standards for a group of programming languages [15]. One of the standards is the SEI CERT Oracle Coding Standard for Java [17]. To facilitate the application of the Java secure coding rules to the software development practice, the system titled Secure Coding Assistant, that is available at http://benw40 8701.github.io/SecureCodingAssistant/, was developed at California State University Sacramento, in the Eclipse development environment, to automate the early detection for a subset of the Java secure coding rules during coding time [19, 20]. This system can help Java programmers significantly reduce security vulnerabilities in their code caused by the violations of secure coding rules. Since it is also a common knowledge that many other software defects can lead to security vulnerabilities, it is highly desirable to extend Secure Coding Assistant not only for detecting violations of secure coding rules but also for assisting programmers to detect, locate and remove errors in their code. It is also highly desirable to perform all these detections during coding time. To this end, efforts have been taken at California State University Sacramento in recent years through a number of projects.

Inspired by Bertrand Meyer's Design by Contract methodology effective for developing robust and reliable software products [10], one project was conducted to enhance Secure Coding Assistant by using Design by Contract [6]. This enhancement was implemented by integrating Secure Coding Assistant with an existing open-source software tool Cofoja that provides the functionality for Design by Contract [7]. Programmers using this enhanced system are advised to provide design contracts, i.e., logic assertions of pre-conditions and post-conditions, for the methods defined in their code. The programmers defined design contracts can then be automatically and dynamically checked to help programmers detect and remove defects in their code. This project demonstrated that detecting violations of secure coding rules and detecting other software defects can be automated or semi-automated in the same system. However, the program structures covered then needed to be extended, and how to help programmers detect and finally locate errors in lengthy code remained a challenge.

Another project separated from Secure Coding Assistant was conducted to experiment the combination of Design by Contract and Programming Logic [1]. A tool called Subcontractor for Java was built upon an open-source tool OpenJML for Design by Contract [13]. Programmers provided design contracts are automatically checked at dynamic time of their program execution. In addition, for program structures of if-then-statements and while-loop statements, this tool can generate sub-design contracts from the programmers provided design contracts. The automatic generation of sub-design contracts is based on the inference rules for the if-then-else statement and the while-loop statement in programming logic [11, 14]. The tool generated sub-design contracts can be automatically inserted into the proper places in code for dynamic checking. This tool (i.e., Subcontractor) demonstrated the feasibility of combining Design by Contract and Programming Logic. By decomposing an original large program verification problem into smaller program verification problems, this tool can facilitate the process of detecting, locating and removing of code defects. However, this tool development was heavily relying on OpenJML, and making the detection of secure coding rule violations and the detection of software defects by the demonstrated combination both available in the Java development environment such as Eclipse remained a challenge.

Building upon the experiences of the above-mentioned two projects, a new project has been designed and conducted to address the challenges for the extension of Secure Coding Assistant. As a result, this paper presents an enhancement to Secure Coding Assistant by a combination of Design by Contract and Programming Logic. Java programmers using this enhanced system are advised to provide their design contracts of logic assertions for three program structures, i.e., methods, if-then-else statements and while-loop statements. For each method or if-then-else statement, its design contract is formed by its pre-condition and post-condition. For each while-loop statement, its design contract consists of its pre-condition, post-condition and loop-invariant. The programmers-defined design contracts can be automatically inserted into proper places in the code and checked at dynamic time. Based on the inference rules for the if-then-else statement and the while-loop statement in programming logic [11, 14], sub-design contracts can be automatically generated from the programmers-defined design contracts. The system-generated sub-design contracts can also be automatically inserted into code and dynamically checked. As an additional feature, based on the assignment axiom and the inference rule for the sequence statement [11, 14], the weakest pre-conditions of certain assignment sequences can be automatically generated from the post-conditions of the sequences, enabling programmers to statically analyze the correctness of the corresponding design contracts they defined. As for the implementation of this project, all functionalities extended are implemented mainly by augmenting source code and in a way cohesive with the implementation of the original system without using any open-source tools for Design by Contract. Compared with previous versions of the system, the current Secure Coding Assistant with the enhancement presented can better help programmers for the early detections not only for violations of secure coding rules but also for errors in code. All the detections are performed in unison with the coding process of software development to pursue software security.

2 Background and Related Work

2.1 Design by Contract

Bertrand Meyer's Design by Contract methodology was the first implemented in the programming language *Eiffel* [9, 10]. With the *Eiffel* programming environment, programmers are encouraged to specify design contracts during coding development. Each design contract is formed by logic assertions on obligations and benefits. The clients or users of the software functionality must meet the obligations specified. The suppliers of the software functionality must provide the benefits specified if the obligations are met. The obligations and benefits are usually specified in terms of pre-conditions, post-conditions, and invariants. The *Eiffel* language provides programmers with key words to specify design contracts. The system automatically checks the programmers-defined design contracts at dynamic time of their program execution. This methodology significantly improves the robustness and reliability of software products. In addition, in *Eiffel* system, design contracts are used to create software documents that are semantically consistent with the software products finally delivered.

Java does not directly support Design by Contract as a built-in feature. However, there are several tools that are developed to provide Design by Contract for the Java programming language, such as Cofoja [7], iContract [5], Jass [3], and OpenJML [13].

2.2 Secure Coding Assistant

There are several existing tools that can help programmer to detect vulnerability of their code. However, most of these tools are close source static analysis tools and do not provide early detection of security vulnerabilities in the code. The first version of Secure Coding Assistant developed at California State University Sacramento is an open source tool based on early detection for static analysis of software security vulnerabilities [19, 20]. The design of this tool is inspired and based on a subset of the security rules for Java developed by SEI CERT at Carnegie Mellon University. Secure Coding Assistant automates the detection of violations of security rules during the time of coding development. Since its birth, the Secure Coding Assistant system has gone through a series of updates, including the automation of a subset of the SEI CERT securer coding rules for the C programming language [12, 16].

One of the important updates is the enhancement by the Design by Contract methodology. The open source tool Cofoja is integrated with Secure Coding Assistant to provide contract programming for Java programmers using the Eclipse development environment [6, 7]. This enhancement demonstrated that detecting violations of secure coding rules and detecting other software defects can be automated or semi-automated in the same system. However, this enhancement heavily relied on the open source tool Cofoja. Only the program structure of methods was covered. Furthermore, how to help programmers detect and finally locate errors in lengthy code was not addressed.

2.3 Subcontractor

Subcontractor is a tool that combines Design by Contract and Programming Logic [1]. It was implemented based on the open source tool OpenJML. In addition to dynamically checking programmers-defined design contracts, Subcontractor can help programmers locate logic errors in their large pieces of code by automatically generating and examining sub-design contracts. The sub-design contracts are generated based on inference rules for if-then-else statements and while-loop statements in programming logic [11, 14]. Subcontractor demonstrated the feasibility of combining Design by Contract and Programming Logic. This combination enables the decomposition of bigger program verification problems into smaller ones using the inference rules. However, this tool development was heavily relying on OpenJML, and was yet to support the detection of secure coding rule violations.

3   The Enhancement to Secure Coding Assistant

3.1   Goal

This enhancement presented is to enable both detection of secure coding rule violations and detection of program errors by using a combination of Design by Contract and Programming Logic. The enhancement aims at providing the following:

- Advising programmers to provide their design contracts of logic assertions for three program structures, i.e., methods, if-then-else statements and while-loop statements. For each method or if-then-else statement, its design contract is formed by its pre-condition and post-condition. For each while-loop statement, its design contract consists of its pre-condition, post-condition and loop-invariant.
- Generating automatically sub-design contracts for if-then-else statements and the while-loop statements based on inference rules in programming logic [11, 14].
- Checking dynamically programmers-defined design contracts and system-generated sub-design contracts.
- Generating automatically the weakest pre-conditions of certain assignment sequences from the post-conditions of the sequences based on the assignment axioms and the inference rule for sequence statements in programming logic [11, 14], to help programmers statically analyze their design contracts.

3.2   Functionality

The tool can detect if programmers do not provide design contracts, i.e., pre-conditions, post-conditions and/or invariants to the methods, if-then-else statements or while-loop statements in the source code. If the design contracts for those structures are missing in the code, a marker will be created by the system and a dialog box will be prompted to advise programmers to specify their design contracts in the source code.

Functionality for Methods. As shown in Fig. 1, a method structure missing its design contract is detected and a dialog box is prompted to advise programmers to provide their pre-condition and post-condition for the method. The syntactic format for writing pre-condition starts with //@Precodition and is followed by the programmer-provided pre-condition. The syntactic format for writing post-condition is similar. When clicking on "Add precondition and postcondition," the insertion of user-provided pre-condition and post-condition is performed.

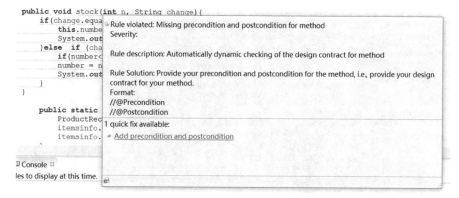

Fig. 1. Advice for the definition of design contracts for methods.

As shown in Fig. 2, after the programmer provides the pre-condition and post-condition for the method, the programmer can elect to automatically perform the dynamic checking of design contracts for the method. The system will generate code for overall pre-condition and post-condition checking based on the assertions given.

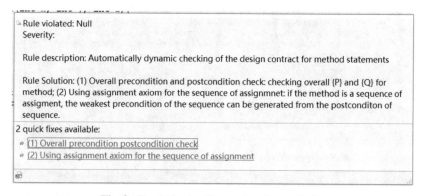

Fig. 2. Dynamic checking options for methods.

Functionality for If-Then-Else Statements. Figure 3 shows the dialog box created by the system when an if-then-else statement is detected missing its pre-condition and post-condition. The programmer is advised to provide the design contract for the statement.

The syntactic format for writing pre-condition of the if-then-else statement starts with //@Precodition and is followed by the programmer-provided pre-condition. However, more specific advice on writing the post-condition assertion is provided in the dialog. Programmers need to provide the data type of the variable if they need to use the key word *old_* followed by the variable name, to store the old value of the variable. The logic assertion before "or" is for the truth branch of the statement, and the assertion after "or" is for the false branch. When clicking on "Add precondition and postcondition," the insertion of user-provided pre-condition and post-condition is performed.

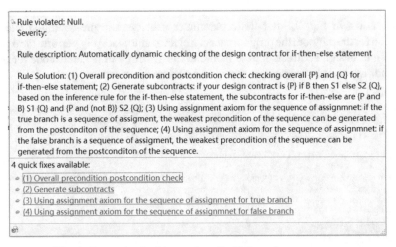

```
public void stock(int n, String change){
    if(change.equals("purchase")){
        this.n
        System
    }else if
        if(num
        number
        System
        }
}

    public sta
        Produc
        itemsi
        itemsi
    }
```

Rule violated: Missing precondition and postcondition for if-then-else statement
Severity:

Rule description: Automatically dynamic checking of the design contract for if-then-else statement

Rule Solution: Provide your precondition and postcondition for the if-then-else statement i.e., provide your design contract for your if-then-else statement.
Format:
//@Precondition
//@Postcondition type variable1 == old_variable2 + variale3(true condition) or type variable1 == old_variable2 + variale3(false condition);

1 quick fix available:
- Add precondition and postcondition

Fig. 3. Advice for the definition of design contracts for if-then-else statements.

Rule violated: Null.
Severity:

Rule description: Automatically dynamic checking of the design contract for if-then-else statement

Rule Solution: (1) Overall precondition and postcondition check: checking overall {P} and {Q} for if-then-else statement; (2) Generate subcontracts: if your design contract is {P} if B then S1 else S2 {Q}, based on the inference rule for the if-then-else statement, the subcontracts for if-then-else are {P and B} S1 {Q} and {P and (not B)} S2 {Q}; (3) Using assignment axiom for the sequence of assignmnet: if the true branch is a sequence of assigment, the weakest precondition of the sequence can be generated from the postconditon of the sequence; (4) Using assignment axiom for the sequence of assignmnet: if the false branch is a sequence of assigment, the weakest precondition of the sequence can be generated from the postconditon of the sequence.

4 quick fixes available:
- (1) Overall precondition postcondition check
- (2) Generate subcontracts
- (3) Using assignment axiom for the sequence of assignment for true branch
- (4) Using assignment axiom for the sequence of assignmnet for false branch

Fig. 4. Dynamic checking options for if-then-else statements.

As shown in Fig. 4, after the programmer provides the pre-condition and post-condition for the if-then-else statement, the programmer can use "Overall precondition and postcondition check" to dynamically check the design contracts provided by the programmer. If the programmer wants to further analyze the code to locate errors, electing "Generate subcontracts" will lead to the automatic generation of sub-design contracts, based on the inference rule shown below for the if-then-else statement in programming logic [11, 14].

$$\frac{\{P \wedge B\}\, S1\, \{Q\}, \{P \wedge (not\, B)\}\, S2\, \{Q\}}{\{P\}\, if\ B\ them\ S1\ else\ S2\, \{Q\}}$$

In the rule above, {P} and {Q} indicate the pre-condition and post-condition given by the programmer. The two sub-design contracts generated by the system for if-then-else statement are $\{P \wedge B\}\, S1\, \{Q\}$ and $\{P \wedge (not\, B)\}\, S2\, \{Q\}$. B is the condition of the if-then-else statement.

After the programmer elects "Overall precondition and postcondition check" and/or "Generate subcontracts", the system will generate code for the overall pre-condition and post-condition checking based on the assertions given by the programmer, and also generate code for the dynamic checking of the sub-design contracts created.

Functionality for While-Loop Statements. For a while-loop statement, there is an additional logic assertion called loop-invariant in the design contract. When a while-loop statement is detected missing its pre-condition, post-condition, and loop-invariant, a dialog box as shown in Fig. 5 is prompted, to advise the programmer to define the design contract for the statement. The syntactic format for writing the design contracts for while-loop statements is similar to that for methods and if-then-else statement. When clicking on "Add precondition, postcondition and invariant," the programmer-provided pre-condition, post-condition and loop-invariant are inserted to the code.

```
public void cal(int c){
    int x = c;
    int y = 0;
    System.out.println("Before: x is " + x + ", y is " + y);
    while(x>0){
        x = x
        y = y       Rule violated: Missing precondition, postcondition and invariant for while-loop statement
    }              Severity:

}
    public static    Rule description: Automatically dynamic checking of the design contract for while-loop statement

    methodtest    Rule Solution: Provide your precondition, postcondition and invariant for the while-loop statement.
    test.cal(6    i.e., provide your design contract for your while-loop statement.
    }              Format:
                   //@Precondition
}                  //@Postcondition
                   //@Inv

                   1 quick fix available:
                    Add precondition, postcondition and invariant
```

Fig. 5. Advice for the definition of design contracts for while-loop statements.

As shown in Fig. 6, after the programmer provides the pre-condition, post-condition and loop-invariant for the while-loop statement, the programmer can use "Overall pre-condition and postcondition check" to dynamically check the design contracts provided by the programmer. If the programmer wants to further analyze the code to locate errors, electing "Generate subcontracts" will lead to the automatic generation of sub-design contracts, based on the inference rule shown below for the while-loop statement in programming logic [11, 14].

$$\frac{P \Rightarrow Inv, \; \{Inv \wedge B\} \, S \, \{Inv\}, \; Inv \wedge (not\,B) \Rightarrow Q}{\{P\} \; while \; B \; do \; S \; \{Q\}}$$

In the rule above, {P}, {Q}, and Inv indicate the pre-condition, post-condition, and the loop-invariant provided by the programmer. The three sub-design contracts generated by the system for while-loop statement are $P \Rightarrow Inv$, $\{Inv \wedge B\} \, S \, \{Inv\}$, and $Inv \wedge (not\,B) \Rightarrow Q$. $P \Rightarrow Inv$ must be satisfied before the execution enters the loop, $\{Inv \wedge B\} \, S \, \{Inv\}$ ensures the preservation of the loop-invariant property after each time of executing the loop body. $Inv \wedge (not\,B) \Rightarrow Q$ must be satisfied right after the execution of the entire loop statement.

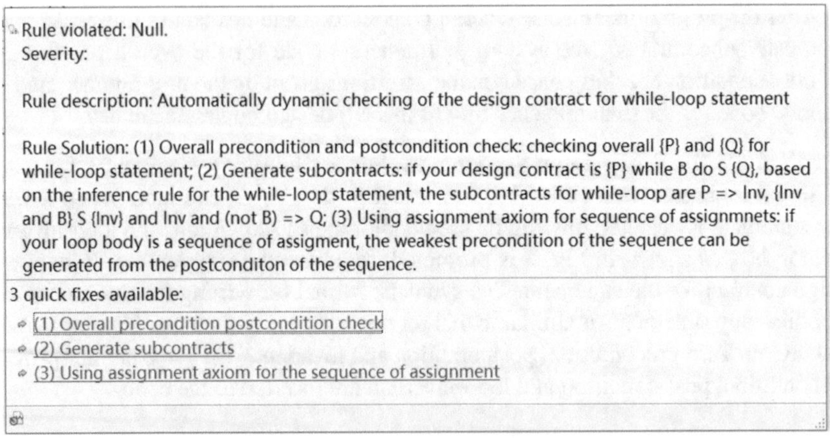

> ⟲ Rule violated: Null.
> Severity:
>
> Rule description: Automatically dynamic checking of the design contract for while-loop statement
>
> Rule Solution: (1) Overall precondition and postcondition check: checking overall {P} and {Q} for while-loop statement; (2) Generate subcontracts: if your design contract is {P} while B do S {Q}, based on the inference rule for the while-loop statement, the subcontracts for while-loop are P => Inv, {Inv and B} S {Inv} and Inv and (not B) => Q; (3) Using assignment axiom for sequence of assignmnets: if your loop body is a sequence of assigment, the weakest precondition of the sequence can be generated from the postconditon of the sequence.
>
> 3 quick fixes available:
> ⟐ (1) Overall precondition postcondition check
> ⟐ (2) Generate subcontracts
> ⟐ (3) Using assignment axiom for the sequence of assignment
>
> 🐿

Fig. 6. Dynamic checking options for while-loop statements.

Functionality for a Sequence of Assignments. If the body of a method, a branch of an if-then-else statement, or a body of a while-loop statement is a sequence of assignments, there is an additional option called "Using assignment axiom for sequence of assignment", available in the dialog boxes as shown in Fig. 2, Fig. 4 and Fig. 6. This option will not be available for other types of code structures. The axiom with backward method for assignments and the inference rule for the sequence statements are used to generate the weakest pre-condition of the sequence from the post-condition of the sequence. Below are the axiom for assignments [11, 14] and the inference rule of sequence of statements [11, 14] used in this process:

$$\{P\}v := E\{Q\}$$

$$\frac{\{P\}\,S1\,\{R\},\ \{R\}\,S2\,\{Q\}}{\{P\}\,S1;\ S2\,\{Q\}}$$

3.3   Workflow

As a system for early detection, Secure Coding Assistant can provide immediate feedback to programmers. It runs in the background to monitor code changes and detect rule conflicts [19, 20]. Abstract Syntax Tree (AST) is used to represent the structure of the source code for analysis. When a rule violation is detected, a marker is created where the violation takes place. Any subsequent code changes will clear all existing markers and trigger a new round of AST node traversal [6].

Figure 7 shows the workflow of functionalities extended to the Secure Coding Assistant. Specifically, this enhancement to Secure Coding Assistant inherits the previous design for all monitoring in the background and AST node traversal. The function *violated()* in Secure Coding Assistant travels across AST nodes. If there is a method, an if-then-else statement or a while-loop statement, a marker will be created. If any design contract, i.e., pre-condition, post-condition and/or loop-invariant, is detected missing, the system will advise programmers to define the design contract in the form of logic assertions. After logic assertions are provided, the function *getSolutions()* of Secure Coding Assistant will be called to provide programmers with the option of inserting code for dynamic checking of programmer-defined overall design contracts. If the execution of the program does not lead to expected results, and if the programmer wants to further analyze the code to locate errors, the system provides another option, called "Generate subcontracts", to generate sub-design contracts for if-the-else statements and while-loop statements. The code for this dynamic checking will be automatically inserted into the programmer's code. When the augmented code is executed in the Eclipse environment, the programmer can analyze the result of the dynamic checking of all the assertions. If the body of a method, a branch of an if-then-else statement, or a body of a while-loop statement is a sequence of assignments, the function *getSolutions()* is called to provide programmers with the option of generating the weakest pre-condition of the sequence from the post-condition of the sequence.

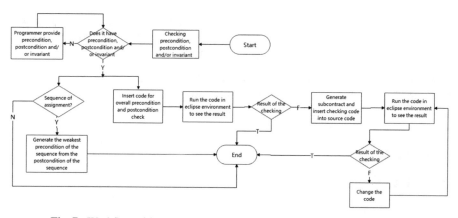

Fig. 7. Workflow of functionalities extended to the Secure Coding Assistant.

3.4 Implementation

To implement Design by Contract, the key words *@Precondition*, *@Postcondition* and *@Inv* are defined, and programmer can use comments to write logic assertions to the source code if needed. Functions in Java's *IWorkbenchWindow* interface are used to get active page for the workbench window and the Java function *getPath()* is used to get the path of the source code. Eclipse JDT *ASTParser* is used to parse Java source files by using the path and the Java function *getCommentList()* from the *CompilationUnit* interface is used to get comments. The Java function *accept()* is used to make the comments accessible. The Java function *visit()* from ASTVisitor class is overridden to store pre-condition, post-condition and/or loop-invariant into the Java *Map* structure. The function *violated()* in Secure Coding Assistant is overridden to identify methods, if-then-else statements or while-loop statements in the source code. If one of these structures is identified, a marker will be created in that location.

To insert code for dynamic checking of the overall pre-condition and post-condition checking, and to insert code for dynamic checking of the system-generated sub-design contracts, source code needed to be edited. The Java function *getASTRewrite()* from *ASTRewrite* class is used in Secure Coding Assistant function *getSolutions()*, which is rewritten to insert the generated code into the proper places in the source code. The *getASTRewrite()* returns an ASTRewrite instance from which this ListRewriter was created [2]. The Java functions *insertFirst(), insertLast(), insertAfter()* and *insertBefore()* are used to insert code for dynamic checking of logic assertions to their corresponding location.

More design and implementation details for the enhancement presented are available from the project report [8].

4 Examples

To evaluate the enhancement presented in this paper, we run the system for several examples. Below please find illustrative results for an if-then-else statement, a while-loop statement, and a sequence of assignments statements.

Example for an If-Then-Else Statement. An example of an if-then-else statement is shown in Fig. 8. Following the system advice, the programmer provides the pre-condition and the post-condition for this if-then-else statement.

```
public class Account {
    private int balance;

    public Account() {
        this.balance = 0;
    }

    //@Precondition amount > 0;                    Programmer-defined design contract
    //@Postcondition int balance == old_balance + amount or int balance == old_balance - amount;
    public void test(int amount , String operation){
        if(operation.equals("deposit")){
            balance= amount + balance;
            System.out.println("balance after deposit amount "+ amount +" is: "+balance);
        }else  if (operation.equals("withdraw")){
            if(balance<amount) throw new RuntimeException();
            balance = amount - balance;
            System.out.println("balance after withdraw amount "+amount+ " is: "+ balance);
        }
    }

    public static void main(String[] args) {
        Account SampleAccount = new Account();
        SampleAccount.balance=1000;
        SampleAccount.test(50, "withdraw");
    }
}
```

Fig. 8.   Source code with programmer-defined design contract for an if-then-else.

If the programmer clicks the option called "Overall precondition and postcondition check" as shown in Fig. 4, the system will automatically generate code for dynamical checking of the design contract. Figure 9 shows that this code along with system-generated comments is inserted into the original code. When the programmer executes the augmented code in Eclipse, the result of the overall pre-condition and post-condition checking is given in Fig. 10. The result shows that the pre-condition is true, the post-condition is false, and the result of the program execution is −950 that is not correct. This means that there is some error either in the code or in the design contract.

```
public class Account {
    private int balance;

    public Account(){
        this.balance = 0;
    }

    //@Precondition amount > 0;                    Programmer-defined design contract
    //@Postcondition int balance == old_balance + amount or int balance == old_balance - amount;
    public void test(int amount , String operation){
        //Insert code:
        boolean OVERALLPRECONDITION = amount > 0;
        int old_balance = balance;                    System-generated code for
        System.err.println("Overall Precondition is "+OVERALLPRECONDITION);  dynamic checking of the de-
        //end
        if(operation.equals("deposit")){             sign contract
            balance= amount + balance;
            System.out.println("balance after deposit amount "+ amount +" is: "+balance);
        }else if (operation.equals("withdraw")){
            if(balance<amount) throw new RuntimeException();
            balance = amount - balance;
            System.out.println("balance after withdraw amount "+amount+" is: "+ balance);
        }
        //Insert code:
        boolean OVERALLPOSTCONDITION = balance == old_balance + amount || balance == old_balance - amount;
        System.err.println("Overall Postcondition is "+OVERALLPOSTCONDITION);
        //end
    }
    public static void main(String[] args) {
        Account SampleAccount = new Account();
        SampleAccount.balance=1000;
        SampleAccount.test(50, "withdraw");
    }
```

Fig. 9. Code with pre-condition and post-condition checking for an if-then-else.

```
Tasks  Console ⊠
<terminated> Account [Java Application] D:\java\jdk1.8.0_65\bin\javaw.exe
Overall Precondition is true
Overall Postcondition is false
balance after withdraw amount 50 is: -950
```

Fig. 10. Result of pre-condition and post-condition checking for an if-then-else.

To locate the error, the programmer can click the option called "Generate subcontracts" as shown in Fig. 4. The system will then automatically generate the code for the dynamic checking of sub-design contracts based on the inference rule for if-then-else statements. This code along with related comments generated by the system is inserted into the original code. Figure 11 presents the original code and the inserted code for dynamic checking of the system-generated sub-design contracts. Due to the space limit, the inserted code for the dynamic checking of the programmer-provided design contract is removed from Fig. 11 in this paper. Figure 12 shows that actually the false branch of this if-then-else statement is executed for the dynamic checking of system-generated sub-design contract. As the result, its pre-condition is true and its post-condition is false. This means that the dynamic checking of the sub-design contract helps the programmer locate the error in the false branch of this if-then-else statement. The programmer needs to analyze the design contract and the false branch code to remove the defect. Figure 13 shows the expected correct result of the dynamic checking of this false branch after the programmer changes the line of the code "*balance = amount – balance*" to "*balance = balance – amount*".

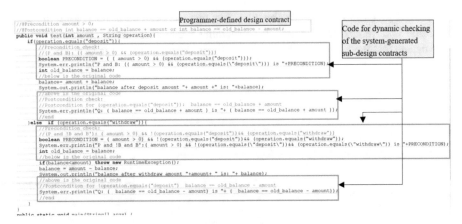

Fig. 11. An if-then-else statement with sub-design contracts.

```
P and !B and B':( amount > 0) && !(operation.equals("deposit"))&& (operation.equals("withdraw")) is true
Q: ( balance == old_balance - amount) is false
balance after withdraw amount 50 is: -950
```

Fig. 12. Result of sub-design contract checking for an if-then-else.

```
balance after withdraw amount 50 is: 950
P and !B and B':( amount > 0) && !(operation.equals("deposit"))&& (operation.equals("withdraw")) is true
Q: ( balance == old_balance - amount) is true
```

Fig. 13. Result of sub-design contract checking for an if-then-else after the code is changed.

Example for a While-Loop Statement. An example of a while-loop statement is shown in Fig. 14. Following the system advice, the programmer provides the pre-condition, the post-condition, the loop-invariant for this while-loop statement.

```
public class Count {
    private static int n = 5;

    //@Precondition n>0 && k==n && mul==0;
    //@Postcondition mul==m*n;
    //@Inv mul==(n-k)*m && k>=0   Programmer-defined design contract
    public int multi(int  n, int m) {
        int mul = 0;
        int k = n;
        while(k>0){
            mul = mul - m;
            k = k - 1;
        }
        return mul;
    }
    public static void main(String[] args) {

        Count count = new Count();
        System.out.println("Product: "+count.multi(5,6));
    }
```

Fig. 14. Java source code with programmer-defined design contracts for a while-loop.

If the programmer clicks the option called "Overall precondition and postcondition check" as shown in Fig. 6, the system will automatically generate code for dynamical checking of the design contract. Similar to the above-given example for if-then-else statement, this code along with system-generated comments is inserted into the original code. When the programmer executes the augmented code in Eclipse, the result of the overall pre-condition and post-condition checking is given in Fig. 15. The dynamic checking result shows that the pre-condition is true, the post-condition is false, and the result of the program execution is −30 that is not the correct one. This means there is some error either in the design contract or in the code.

```
Overall Precondition is true
Overall Postcondition is false
Product: -30
```

Fig. 15. Result of pre-condition and post-condition checking for a while-loop.

To locate the error, the programmer can click the option called "Generate subcontracts" as shown in Fig. 6. The system will then automatically generate the code for the dynamic checking of sub-design contracts based on the inference rule for while-loop statements. This code along with related comments generated by the system is inserted into the original code. Figure 16 presents the original code and the inserted code for dynamic checking of the system-generated sub-design contracts. Due to the space limit, the inserted code for the dynamic checking of the programmer-provided design contract is removed from Fig. 16. Figure 17 shows that $P \Rightarrow Inv$ is true, $\{Inv \wedge B\}S\{Inv\}$ is not satisfied for every time of executing the loop body, and $Inv \wedge (not\ B) \Rightarrow Q$ is true. This means the dynamic checking of the sub-design contract helps the programmer locate the error in this loop body. The programmer needs to analyze the design contract and the code for the loop body to remove the defect. Figure 18 shows the expected correct result of the dynamic checking of this loop statement after the programmer changes the line of the code "$mul = mul - m$" to "$mul = mul + m$".

Example for a Sequence of Assignments. Figure 19 shows an example of using the assignment axiom and the inference rule for the sequence of statements to automatically generate the weakest pre-condition of an assignment sequence from the given post-condition of the sequence. The backward method associated with the assignment axiom is used in the process. For this example, the generated weakest pre-condition is equivalent to the programmer-defined pre-condition. In general, there should be an implication relationship from the programmer-defined pre-condition to the system-generated weakest pre-condition. If this logic relationship is not satisfied, the programmer needs to analyze and then correct the pre-condition, post-condition, and/or code. This analysis and correction are performed statically.

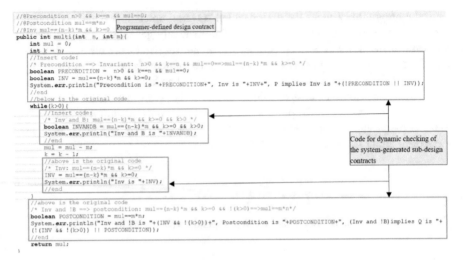

```
//@Precondition n>0 && k==n && mul==0;
//@Postcondition mul==m*n;
//@Inv mul==(n-k)*m && k>=0    Programmer-defined design contract
public int multi(int n, int m){
    int mul = 0;
    int k = n;
    //Insert code:
    /* Precondition ==> Invariant:  n>0 && k==n && mul==0==>mul==(n-k)*m && k>=0 */
    boolean PRECONDITION =  n>0 && k==n && mul==0;
    boolean INV = mul==(n-k)*m && k>=0;
    System.err.println("Precondition is "+PRECONDITION+", Inv is "+INV+", P implies Inv is "+(!PRECONDITION || INV));
    //end
    //below is the original code
    while(k>0){
        //Insert code:
        /* Inv and B: mul==(n-k)*m && k>=0 && k>0 */
        boolean INVANDB = mul==(n-k)*m && k>=0 && k>0;
        System.err.println("Inv and B is "+INVANDB);
        //end
        mul = mul - m;
        k = k - 1;
        //above is the original code
        /* Inv: mul==(n-k)*m && k>=0 */
        INV = mul==(n-k)*m && k>=0;
        System.err.println("Inv is "+INV);
        //end
    }
    //above is the original code
    /* Inv and !B ==> postcondition: mul==(n-k)*m && k>=0 && !(k>0)==>mul==m*n*/
    boolean POSTCONDITION = mul==m*n;
    System.err.println("Inv and !B is "+(INV && !(k>0))+", Postcondition is "+POSTCONDITION+", (Inv and !B)implies Q is "+
(!(INV && !(k>0)) || POSTCONDITION));
    //end
    return mul;
}
```

Code for dynamic checking of the system-generated sub-design contracts

Fig. 16. A while-loop statement with sub-design contracts.

```
Precondition is true, Inv is true, P implies Inv is true
Inv and B is true
Inv is false
Inv and B is false
Inv is false
Inv and B is false
Inv is false
Inv and B is false
Inv is false
Inv and B is false
Inv is false
Inv and !B is false, Postcondition is false, (Inv and !B)implies Q is true
Product: -30
```

Fig. 17. Result of sub-design contract checking for a while-loop.

```
Precondition is true, Inv is true, P implies Inv is true
Inv and B is true
Inv is true
Inv and B is true
Inv is true
Inv and B is true
Inv is true
Inv and B is true
Inv is true
Inv and B is true
Inv is true
Inv and !B is true, Postcondition is true, (Inv and !B)implies Q is true
Product: 30
```

Fig. 18. Result of sub-design contract checking for a while-loop after the code is changed.

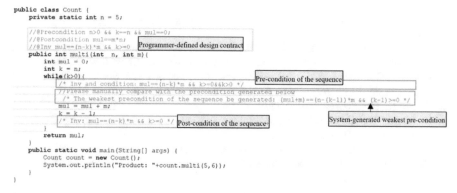

Fig. 19. Result of using assignment axiom and inference rule of sequence statements for a sequence of assignments.

5 Conclusion and Future Work

The enhanced Secure Coding Assistant supports early detection of secure coding rule violations as defined by SET CERT, and the early detection of code defects based on a combination of Design by Contract and Programming Logic. It can help programmers improve software security by removing potential security vulnerabilities during coding. The future work will focus on the nested code structures, such as nested while-loop statements and if-then-else statements. The challenge is to repeatedly or recursively generate sub-design contracts for nested code structures based on inference rules.

References

1. Aldausari, N., Zhang, C., Dai. J.: Combining design by contract and inference rules of programming logic towards software reliability. In: Proceedings of SECRYPT 2018 (2018)
2. ASTRewrite.   https://www.ibm.com/support/knowledgecenter/ko/SSZHNR_1.0.0/org.ecl ipse.jdt.doc.isv/reference/api/org/eclipse/jdt/core/dom/rewrite/ListRewrite.html
3. Bartetzko, D., Fischer, C., Möller, M., Wehrheim, H.: Jass — Java with assertions. Electron. Notes Theor. Comput. Sci. **55**(2), 103–117 (2001)
4. Cybercrime Facts and Statistics. https://1c7fab3im83f5gqiow2qqs2k-wpengine.netdna-ssl. com/wp-content/uploads/2021/01/Cyberwarfare-2021-Report.pdf
5. Kramer, R.: iContract - the Java(tm) design by contract(tm) tool. In: Proceedings of the Technology of Object-Oriented Languages and Systems (1998)
6. Li, C., Dai. J., Zhang, C.: Enhancing secure coding assistant with error correction and contract programming. In: Proceedings of the National Cyber Summit, 6–8 June 2017 (2017)
7. Le, N.M.: Cofoja github page. http://github.com/nhatminhle/cofoja
8. Liang, W.: Combining design by contract and programming logic to enhance secure coding assistant system. MS Project Report, California State University, Sacramento, May 2021
9. Meyer, B.: Eiffel: a language for software engineering. Technical Report TR-CS-85-19 University of California, Santa Barbara (1985)
10. Meyer, B.: Applying 'design by contract.' Computer **25**(10), 40–51 (1992). https://doi.org/ 10.1109/2.161279

11. Meyer, B.: Introduction to the Theory of Programming Languages. Prentice Hall, Hoboken (1990)
12. Melnik, V., Dai, J., Zhang, C., White, B.: Enforcing secure coding rules for the C programming language using the eclipse development environment. In: Choo, K.-K.R., Morris, T.H., Peterson, G.L. (eds.) NCS 2019. AISC, vol. 1055, pp. 140–152. Springer, Cham (2020). https://doi.org/10.1007/978-3-030-31239-8_12
13. OpenJML. http://www.openjml.org/
14. Slonneger, K., Kurtz, B.L.: Formal Syntax and Semantics of Programming Languages. Addison Wesley, Boston (1995)
15. SEI CERT Coding Standards. https://wiki.sei.cmu.edu/confluence/display/seccode
16. SEI CERT C Coding Standard. https://wiki.sei.cmu.edu/confluence/display/c/SEI+CERT+C+Coding+Standard
17. SEI CERT Oracle Coding Standard for Java. https://wiki.sei.cmu.edu/confluence/display/java/SEI+CERT+Oracle+Coding+Standard+for+Java
18. The Hidden Costs of Cybercrime. https://www.mcafee.com/enterprise/en-us/assets/reports/rp-hidden-costs-of-cybercrime.pdf
19. White, B., Dai. J., Zhang, C.: Secure coding assistant: enforcing secure coding practices using the eclipse development environment. National Cyber Summit (2016)
20. White, B., Dai, J., Zhang, C.: An early detection tool in eclipse to support secure coding practices. Int. J. Inf. Priv. Secur. Integr. **3**(4), 284–309 (2018)

Social Engineering Attacks in Healthcare Systems: A Survey

Christopher Nguyen, Walt Williams, Brandon Didlake, Donte Mitchell,
James McGinnis[✉], and Dipankar Dasgupta

Center for Information Assurance, University of Memphis, Memphis, TN 38152, USA
{cnguyen6,wwllams9,bdidlake,dkmtchl3,jmcgnnis,
dasgupta}@memphis.edu

Abstract. Technology runs much of modern society's daily functions due to how efficient, reliable, and easy it is to access and manage content anywhere at any time. This rapid growth has created an emphasis on cybersecurity to ensure data integrity in today's digital realm and the future to come. Since more industries are relying on technology, cybersecurity is becoming more utilized as the foundation for success for many companies and individuals alike. However, as these new avenues for communication become part of daily life, cyber threats have also become more prevalent. One of these avenues affected includes healthcare telemedicine (Annaswarmy et al. 2020) which during COVID-19 pandemic provides patients with more convenient methods of medical services. To prevent cyber-attacks on these services through social engineering, among several defense techniques, including machine learning (ML), are being researched to mitigate the effects of human error. This paper provides recent social engineering attacks on healthcare systems, devices, and telemedicine services; and highlights the potential of machine learning in defending against social engineering attacks.

Keywords: Cybersecurity · Telehealth · Telemedicine · Healthcare · Internet-of-Things · Machine learning · Social engineering

1   Introduction

Cybersecurity is the practice of protecting digital information and systems from unauthorized personnel. All forms of technology that utilize a connection or network are vulnerable to various cybersecurity threats. (Tunggal 2020). Simple tasks that include creating strong passwords, keeping software up to date, utilizing antivirus programs, and limiting the number of people on a network and monitoring activities can protect users from a wide range of cyber-attacks. The growth of cyber threat activity influences more companies to create better security methods to protect their respective companies' information. One field that is increasingly impacted is healthcare. This paper covers a broad basis of how workforce development comes into place within the telehealth field. We analyze the implications of telehealth security due to its newly produced devices, how those devices are affected by social engineering techniques, and the effectiveness of machine learning techniques for defense.

K.-K. R. Choo et al. (Eds.): NCS 2021, LNNS 310, pp. 141–150, 2022.
https://doi.org/10.1007/978-3-030-84614-5_11

1.1 Cybersecurity Issues in Telehealth

Telehealth provides patients with the needed services through the Internet. Since confidential information is being passed through online communication, this field is at a higher risk due to the type of information that is stored within these health networks. In addition, Telehealth often is equipped with more distinct newly developed technology, newer hardware, and software for end-user services. Due to the sensitive information transmission, cybersecurity practices have become more important in the telehealth field (Pennic 2020). This appears to be the faults of patients and workers not possessing encrypted services on their devices, ultimately leaving them vulnerable to cyber-attacks, including social engineering attacks.

2   Use of IoT in Healthcare

Consumer-grade "Internet of Things" (IoT) devices include different health sensors, Fitbit, insulin pumps and meters, pacemaker, and devices such as the Amazon Echo or even the Google Nest family of devices connected in smart homes to make our lives better by specializing and performing specific tasks such as being a personalized assistant. When such IoT devices (Shirer 2019) work together, the resulting support system can provide endless possibilities and solutions for existing problems society has yet to find a solution to.

Internet of Health Things (IoHT) is essentially IoT devices operating and performing functions in a healthcare environment (Sharma 2020; HealthLeaders 2020). IoHT is a more specialized implementation of IoT devices in healthcare sector in order to provide many benefits, such as remote health monitoring, avenues for treatment delivery, and the integration of different sensors for diagnostics (Econsultancy 2019). IoHT has a significant potential for information gathering, scientists and data analysts are able to collect and dissect different data trends in the health sector, which can lead to new discoveries and solutions pertaining to medicine production, distribution, and patient care.

IoT/IoHT devices are able to work in specialized systems which sometimes utilize Linux kernels (Team, KernelCare 2019). Linux kernel-based devices have an underlying software that controls the hardware, that the end-user never touches or modifies. Once these IoT devices are working together in tandem, they are able to perform tasks in a network cooperatively. This network of IoT devices also, in turn, opens the door for attackers to access the network as most of these IoT devices are utilizing outdated and insecure Linux kernels which are constrained due to power requirements.

One issue that IoHT devices and sensors are currently facing is operating wirelessly with limited battery charging. To conserve the power of these batteries, the IoT devices are natively running in a "low powered mode" to allow for a longer battery lifespan, (AVSystem 2019) which prevents them from performing power-intensive security-based functions. Also, end-users are responsible to ensure IoHT devices are security-enabled, well-maintained, and charged to mitigate the potential for an intruder to tamper the devices.

Like most devices connected to a network, IoHT devices need to be updated regularly to ensure that the latest security and optimization patches are installed. By leaving an unpatched device in use, this may allow cyber attackers to perform malicious actions such as contaminate records or steal private health information, etc.

To achieve maximum benefit of IoT devices, there are typically interconnected to share and process data and control devices appropriately. Moreover, such networks are linked to cloud services to offload patient data for archiving and use by other entities such as insurance company, pharmacies etc. Since each device on the network is capable of interacting and communicating with other entities on the same network, this increases the risk for social engineering-based attacks. Moreover, uploading data to a cloud-based storage can create a bigger attack surface for hackers to exploit.

3   Social Engineering Attacks

Social Engineering attacks are primarily conducted by exploiting the trust and lack of awareness of human users while using the Internet. This type of attack is difficult to defend against due to human operators being manipulated into breaching security procedures that are normally in place to prevent these types of attacks. In addition to the fact that social engineers (hackers) are constantly evolving their attack vectors, this creates an interweaving problem that is difficult to prevent using basic security procedures (Frumento 2018).

While social engineering can be complicated, most common attacks follow a common attack pattern that utilizes multiple techniques called the Social Engineering cycle. The Social Engineering cycle consists of five core steps. This is worth mentioning because of a dogmatic process, and the cycle can start at any point (See Fig. 1).

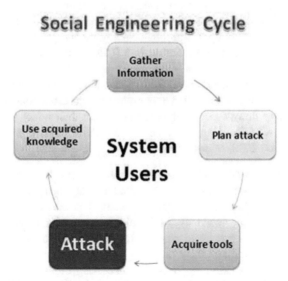

Fig. 1. Illustrates the Social Engineering cycle (Cihodariu 2020)

1. Gather Information: This phase is when the hacker acquires knowledge about their target. Hackers can gain information through public domain sites that contain information about their target or are physically exchanging information pertaining to the target.
2. Plan Attack: This phase is when the hacker starts to process collected information the hacker has about their target and decides what specific information the hacker will be looking for during the attack phase.
3. Acquire Tools: In this phase, the hacker sets up the tools needed to carry out the attack. This process can range from setting up automatic malware injections to creating fake websites/programs that copy the information before distributing it to the victim.
4. Attack: This phase is when the hacker uses their prior preparations and tools to gain to information from their target.
5. Use Acquired Knowledge: In this phase of the plan, the hacker has acquired the information they were seeking and will utilize learned knowledge during the attack to plan the proceeding steps. Learned knowledge can pertain to the psychological behavior of the victim, network infrastructure, and gathered information during the attack attempt.

3.1 Common Social Engineering Attacks

Hackers use a multitude of different techniques that allow them to breach different types of networks. Different techniques hackers often utilize are phishing, vishing, whaling, spear phishing, and impersonation.

Phishing. Phishing is a social engineering attack that happens through the medium of messaging services, which can be text messages or emails. Hackers usually rely on alarmist information, such as relying on feelings of fear, urgency, and want to be helpful for people in need. Making this problem even more complicated is that on average a phishing attack can be successful within 24 h (Cimpanu 2020). In telemedicine, this problem has become exasperated due to the increased growth in the last few years. Using the same tactic of relying on alarmist information, hackers sent out emails in mass with information concerning COVID-19. These emails would often contain links or try to solicit the user into downloading software that would then infect the computer system. Malicious-based attacks have the potential to undermine the future of telemedicine utilization (Clarkson 2020; Lyons 2020).

Vishing. Vishing is a social engineering attack that happens through phone calls. They can either be a pre-recorded message or a person actively trying to convince someone to give away secure information. Vishing is generally received via robocalls directed towards people with vague knowledge of the telemedicine market. While at times legitimate, such as doctors being assigned to treat patients this way, they could also be involved in fraudulent activities utilizing false or stolen credentials that would utilize a patient's records and generate prescriptions to make unnecessary orders. This case is exasperated in the telemedicine market due to the lack of oversight and practicing guidelines being neglected. Due to this neglect, hackers have the ability to utilize doctor's credentials to sham medical companies and government institutions to wire money away from telemedicine services (Knight 2019; Cowart 2019).

Spear Phishing. Spear phishing occurs when one specific person is being targeted for their information. In telemedicine, the situation can escalate quickly should a patient information become compromised. With customer's confidential information, hackers could target services directly relating to the patient's information and convince the patient that there are possible solutions that are actively misleading. Spear phishing also runs the risk of hackers ordering excessive amounts of the treatment product, which could be billed towards either the patient or the insurance company (Knight 2019).

Impersonation. Impersonation is done when a hacker uses false credentials to get information that people would not normally give up casually. In telemedicine, it is much easier because of the lack of physical confirmation required of doctors to have detailed conversations with their patients in order to confirm a prescription. In turn, this gives hackers the ability to make orders relatively unchecked for patients that may not have requested them. The problem is not solved exclusively either by having video conferences since organizations do not have a specific service, they utilize that to verify their legitimacy. Not including the security risk that comes with utilizing third-party sources that were not designed for this type of communication in the beginning (Zoom).

Social Engineering attacks can cause serious damage to the reputation of telemedicine in the long run (Clarkson 2020; Knight 2019; Cowart 2019).

3.2 Social Engineering Attacks in Healthcare Systems

The frequency of data breaches in the healthcare sector have seen an increase in frequency during the past several years (McLeod 2018). Figure 2 shows the number of data breaches involving more than 500 people from 2009–2019. Several high-profile attacks have occurred during this time and caused massive disruption thereby fueling public distrust in secure health care systems. For example, on January 12, 2021, a data breach involving a compromised email resulted in an attack that affected roughly twenty-one thousand customers from Precision Spine Care, a healthcare company in Texas (DHHS). In a breach of 2015 attackers launched a phishing campaign and were able to gain the credentials of an employee with access to private information (Alder 2015). Roughly twenty-five thousand people had their personal healthcare information compromised, such as social security numbers, genders, dates of birth, and health insurance information. In 2019, a reported 95% of healthcare organizations who were targeted by attackers saw email spoofing, a social engineering attack, on their domain (Davis 2019). The role of digital healthcare services is increasing in our present world, and historical trends point to more frequent data breaches in the future. This trend demands the emergence of novel solutions.

During the initial rise of the Covid-19 pandemic, there was an influx of phishing emails that would impersonate government organizations such as the World Health Organization (WHO) that would try to convince users to download software containing information relating to stimulus checks and information about the virus. There were also attempts to have people download links by impersonating upper management to receive new company policy guidelines towards the virus (Lyons 2020).

In 2019 there was a rise in telemedicine scams that were targeting senior citizens using Medicare where they would prescribe multiple different braces to a single patient

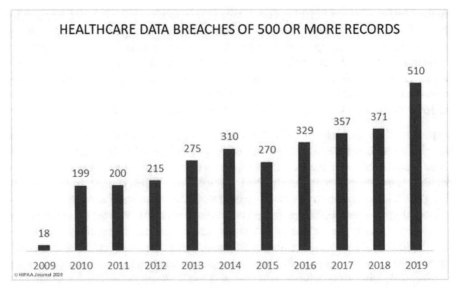

Fig. 2. A bar chart detailing the number of data breaches affecting 500 or more people from 2009–2019 (HIPAA, Healthcare Data Breach Statistics)

using false licenses and sham companies. The problem was amplified by patients not being able to differentiate between calls from medical doctors and hackers claiming to be one. Due to the lack of communication procedures in regular telemedicine calls, a doctor can make the call, take as little or as much time as possible or similar to an in-person visitation. This problem eventually escalated to the point where the US Department of Justice (DOJ) launched an investigation titled "Operation Brace Yourself." (Knight 2019; FBI 2019).

There was a ransomware attack in February 2020 that leaked the data of 657,392 donors from the software company Blackbaud that provides cloud solutions to medical companies (Brown 2020).

The Water Nue campaign was identified by security company Trend Micro that had Business Email Compromise (BEC) from internal Office 365 accounts that takes users to fake versions to steal more login data (HIPAA, Healthcare Data Breach Statistics).

4 Handling Social Engineering Attacks

This section details some effective practices for limiting the damage caused by Social Engineering attacks and makes a case for automated approaches and how they show incredible potential in this fight.

4.1 Tactics for Prevention

Social Engineering can greatly undermine confidence in an organization's security, leading to problems that can greatly affect the timing of accomplishing tasks. Therefore, it is best practice to have annual cybersecurity awareness training and basic guidelines to ensure personnel is always adequately trained and informed on the latest threats. Here are some basic examples to follow (Cihodariu 2020; Johnson 2020):

- Open messages from trusted sources only and if something looks suspicious check the sender receipts. A hacker will often use a similar looking "send address", though it will be slightly misspelled in hard to notice areas.
- Only download software from approved sources and verify their origin point.
- Do not open unchecked attachments since they could install malware onto your computer with the intent of stealing information.
- Do not disable antivirus software.
- Contact the Information Technology (IT) department for updated security guidelines.
- Reduce the number of personal devices on the network to help deduce any foreign connection and centralize the network activity.
- Utilize firewalls and ensure that new application updates are secure due to the risk of Zero-day threats that can compromise unidentified exploits.

4.2 Using Automated Solutions

Cybercriminals attack the weakest link in a cyber chain, and oftentimes the weakest link is people. When it comes to detecting and mitigating social engineering attacks, there have been several effective solutions proposed involving annual employee training and security policies. However, the problem persists that both of these still involve the human element, which is the weakest link. Several automated approaches have been proposed to limit the role of human susceptibility in these attacks as a result. For example, SEDA (Social Engineering Defense Architecture) was one of the earliest automated defense systems for detecting social engineering attacks (Hoeschele 2006). SEDA detects attacks by matching voice signatures from incoming calls with pre-existing voice recordings of employees along with identifiable information such as the employee's name and job title. This matching of voice signatures from a database of employees prevented social engineers from calling a company pretending to be an employee and gaining access to sensitive information. Machine learning (ML) is a technology that has gained incredible popularity in recent years and has been a boon in several research fields. In this section we will briefly cover what ML is and how it helps solve the problem of human error in social engineering attacks.

4.3 Use of Machine Learning to Detect Social Engineering Attacks

In short, machine learning (ML) represents a set of computational methods that are used to automatically acquire information from data (Dasgupta 2020). There exist many ML techniques, Deep Learning is becoming a popular one (Krizhevsky 2012), and AlexNet, ImageNet, TensorFlow are various implementation of deep learning models for image recognition. We feel that machine learning will play a big role in defending against social engineering attacks based on recent evidence. As an example, (Lansley 2020) proposes SEADer++, a machine learning based social engineering detection model. SEADer++

uses a completely automated process to extract features from text-based messages and then proceeds to use ML for the binary classification of the texts. SEADer++ was tested on a synthetic dataset created by Lansley and his team by incorporating a mix of social engineering and "clean" messages with the highest performing model achieving an accuracy of 92.6% on the dataset. Another proposed automated solution for the detection of phishing emails uses a multi-stage approach based on NLP (Natural Language Processing), a set of machine learning solutions for the interpretation and translation of text (Gualberto 2020). The best-performing model achieved a reported accuracy of 100% in their testing set. One important note to consider is that these models perform well inside the same distribution of data they were trained on but have not been tested "in the wild" i.e., with data from different distributions. More research needs to be done in order to gauge the effectiveness of these models in real-world environments.

Models like the two discussed show the promise that ML offers for the detection of social engineering attacks in workplaces. In an industry that contains sensitive and personally identifiable information, like the healthcare industry, it is vital that we limit any data breaches and make these systems as secure as possible. ML's biggest advantage continues to be the automation of detection without the need for humans, thereby limiting the role of human error in social engineering attacks. In this way, ML can help can limit the exploitation of the weakest link in many cyber systems: human beings.

5 Summary

As the world becomes more interconnected and digitalized, a cyber-attack's potential to cause massive disruption to online systems increases. The number of large-scale (>500 affected individuals) data breaches is increasing year after year, and the trend is projected to continue into the foreseeable future. This worrying trend demands the emergence of novel solutions to fight against data theft and data breaches, solutions that help to reduce the impact of the vulnerabilities in cyber systems. In the case of Social Engineering attacks, humans are the primary targets for cybercriminals when launching attacks such as phishing campaigns and impersonation. It may be very difficult to completely eradicate the role of human error to avoid social engineering attacks. Developing a workforce that is aptly trained and informed on the latest cyber threats is crucial to prevent health data breaches. We hope this paper serves as a viable introduction to social engineering attacks in the healthcare space and how automation can help to ameliorate human beings' susceptibility to these attacks. We are optimistic that in the future more solutions, including automated approaches, will be developed for the efficient and effective mitigation of social engineering attacks, especially in industries that hold such valuable and sensitive information as the healthcare industry.

References

Alder, S.: Saint agnes health care hack exposes 25000 HIPAA records (2015). https://www.hipaajournal.com/saint-agnes-healthcare-hack-exposes-25000-hipaa-records-5663/

Annaswarmy, T.M., Verduzco-Gutierrez, M., Frieden, L.: Telemedicine barriers and challenges for persons with disabilities: Covid 19 and beyond, October 2020. https://www.sciencedirect.com/science/article/abs/pii/S1936657420301047?via%3Dihub

AVSystem. IoT standards and protocols guide — protocols of the Internet of Things (2019). www.avsystem.com/blog/iot-protocols-and-standards

Ayoade, G., et al.: Evolving advanced persistent threat detection using provenance graph and metric learning. In: 2020 IEEE Conference on Communications and Network Security (CNS) (2020). https://doi.org/10.1109/cns48642.2020.9162264. https://ieeexplore

Brown, N.: Phishing attacks are targeting healthcare... again (2020). https://www.nextech.com/blog/phishing-attacks-are-targeting-healthcare-again

Cihodariu, M.: What is social engineering: the tactics used to manipulate you (2020). https://heimdalsecurity.com/blog/what-is-social-engineering-tactics/

Cimpanu, C.: Phishing campaigns, from first to last victim, take 21h on average (2020). https://www.zdnet.com/article/phishing-campaigns-from-first-to-last-victim-take-21h-on-average/

Clarkson, K.: Phishing and security risks in telehealth and video communication (2020). https://www.pulsara.com/blog/phishing-and-security-risks-in-telehealth-and-video-communication

Cowart, H.: Telemedicine fraud: how are doctors affected? (2019). https://www.hchlawyers.com/blog/2019/december/telemedicine-fraud-how-are-doctors-affected-/

Dasgupta, D.A.: Machine learning in cybersecurity: a comprehensive survey J. Def. Model. Simul.: Appl. Methodol. Technol., 1–50 (2020). https://doi.org/10.1177/1548512920951275

Davis, J.: Hackers targeting healthcare with social engineering, Email Spoofing (2019). https://healthitsecurity.com/news/hackers-targeting-healthcare-with-social-engineering-email-spoofing

DHHS. DHHS table of healthcare breaches (n.d.). https://ocrportal.hhs.gov/ocr/breach/breach_report.jsf

Econsultancy. 10 examples of the Internet of Things in healthcare, 01 February 2019. https://econsultancy.com/internet-of-things-healthcare/

FBI. Billion-dollar bust, 09 April 2019. https://www.fbi.gov/news/stories/billion-dollar-medicare-fraud-bust-040919

Frumento, E.: Social engineering: an IT security problem doomed to get worse (2018). https://medium.com/our-insights/social-engineering-an-it-security-problem-doomed-to-get-worst-c9429ccf3330

Gualberto, E.D.: The answer is in the text: multi-stage methods for phishing detection based on feature engineering. IEEE Access **8**, 223529–223547 (2020)

HealthLeaders. Telehealth tops the list as physician digital health adoption increases (2020). www.healthleadersmedia.com/innovation/telehealth-tops-list-physician-digital-health-adoption-increases/

HIPAA. Healthcare data breach statistics (n.d.). https://www.hipaajournal.com/healthcare-data-breach-statistics/

HIPAA. Protect healthcare data from phishing (n.d.). https://www.hipaajournal.com/protect-healthcare-data-from-phishing/

Hoeschele, M., Rogers, M.: Detecting social engineering. In: Pollitt, M., Shenoi, S. (eds.) DigitalForensics 2005. ITIFIP, vol. 194, pp. 67–77. Springer, Boston, MA (2006). https://doi.org/10.1007/0-387-31163-7_6

Johnson, C.: 3 warning signs of a telemedicine scam (2020). https://clark.com/scams-rip-offs/telemedicine-scams/

Knight, V.: Phone scammers and 'teledoctors' charged with preying on seniors in fraud case (2019). https://www.npr.org/sections/health-shots/2019/10/07/766517003/phone-scammers-and-teledoctors-charged-with-

Krizhevsky, A.S.: ImageNet classification with deep convolutional neural networks. In: Proceedings of the 25th International Conference on Neural Information Processing Systems, vol. 1, pp. 1097–1105 (2012)

Lansley, M., Mouton, F., Kapetanakis, S., et al.: SEADer++: social engineering attack detection in online environments using machine learning. J. Inf. Telecommun. **4**, 346–362 (2020)

Lyons, K.: Google saw more than 18 million daily malware and phishing emails related to COVID-19 last week (2020). https://www.theverge.com/2020/4/16/21223800/google-mal ware-phishing-covid-19-coronavirus-scam

McLeod, A.: Cyber-analytics: modeling factors associated with healthcare data breaches. Decis. Support Syst. **108**, 57–68 (2018)

Pennic, J.: Telehealth and Cybersecurity: what you should know. HIT Consultant (2020). https://hitconsultant.net/2020/07/22/telehealth-cybersecurity-what-you-should-know/

Sharma, N.: Cloud based healthcare services for telemedicine practices using Internet of Things (2020). http://www.jcreview.com/fulltext/197-1597924280.pdf?1622692915

Shirer, M.: The growth in connected IoT devices is expected to generate 79.4ZB of data in 2025, according to a New IDC Forecast (2019). https://www.businesswire.com/news/home/201906 18005012/en/Growth-Connected-I

Team, KernelCare. IoT devices are in desperate need of live kernel patching (2019). https://blog.kernelcare.com/iot-devices-are-in-desperate-need-of-live-kernel-patching

Tunggal, T.A.: Why is cybersecity important (2020). https://www.upguard.com/blog/cybersecu rity-important

Identifying Anomalous Industrial-Control-System Network Flow Activity Using Cloud Honeypots

Neil C. Rowe[✉], Thuy D. Nguyen, Jeffery T. Dougherty, Matthew C. Bieker, and Darry Pilkington

U.S. Naval Postgraduate School, Monterey, CA 93943, USA
{ncrowe,tdnguyen}@nps.edu, dougherty.jeffrey@gmail.com, matt_bieker@hotmail.com, darry.pilkington@navy.mil

Abstract. This work addressed efficient and effective implementation of honeypots (decoy devices) for industrial control systems in cloud services. The honeypots we investigated simulated control systems of a small electrical-power distribution system. Starting with two honeypot software frameworks called Conpot and GridPot, we increased their deceptiveness by adding new obfuscation techniques, new simulated features of an electric grid, and new control interfaces that mimicked the operator interface for an actual power plant. These deceptions were effective in our first experiments with a standalone honeypot. We then deployed the honeypot at three cloud sites in the U.S. and in Asia. Attacks we observed were mostly similar between the deployments with a few differences. We were concerned that deployment in the cloud could be detected by attackers and discourage them, but we saw no significant differences in activity between the two kinds of deployments; apparently enough systems are managed in the cloud today that such deployment is not suspicious. We conclude that honeypots for industrial control systems using cloud services are an effective tool for collecting attack intelligence.

Keywords: Honeypot · Cloud services · Industrial control systems · Power plants · Electric grid · Cybersecurity · Cyberattack · Deception · Conpot · GridPot · Obfuscation · Adversary

1 Introduction

In recent years, critical-infrastructure systems have become increasingly complex, interdependent, and reliant on computerized industrial control systems (ICS) [1]. As our dependence on these systems has grown, so has the frequency and complexity of cyberattacks against them. During the initial development of ICS systems from the 1950s, little attention was paid to identifying and managing risk from potential security flaws because the systems were not connected to broad networks. This has now changed [2]. However, ICS systems are often constructed with a planned lifetime of 20 to 30 years, so legacy systems with well-known vulnerabilities continue to be used.

© The Author(s), under exclusive license to Springer Nature Switzerland AG 2022
K.-K. R. Choo et al. (Eds.): NCS 2021, LNNS 310, pp. 151–162, 2022.
https://doi.org/10.1007/978-3-030-84614-5_12

This work studied power-grid (electric-grid) systems in particular [3]. These have become major users of ICSs for power generation, transmission, and distribution. While this has increased efficiency, it has also left power grids increasingly vulnerable to cyberattack. An example occurred in December 2015 when hackers linked to the Russian government used the malware "CrashOverride" to cut electrical power service to over 230,000 people in the Ukraine [4].

Due to the difficulties in securing the legacy software used in many power systems, researchers are examining novel ways to improve their security. One method is the deployment of honeypots (decoy devices) to collect attack intelligence. Cloud services would seem a good way to implement honeypots since they can support large numbers of simultaneous deployments efficiently. Many cyberattack campaigns choose targets and methods randomly, and offering many targets permits defenders to see more variety of attacks [5].

Industries are increasingly using cloud services for supervisory control of manufacturing, power plants, and other industrial control systems. Honeypots in the cloud for these purposes may be unconvincing to cyberattackers who are alert because using cloud services for supervisory controls is relatively recent, and detection of cloud services is often easy through lookup of the site owner. Nonetheless, so many cyberattacks are automated today that rarely do humans attackers inspect the characteristics of a site anymore.

To confirm this, we can assess the convincingness of cloud honeypots by experimenting with real cyberattackers. Fortunately, cyberattackers need not be recruited, as putting any site up on the Internet usually attracts attackers within minutes, and at a higher rate for industrial control systems than for traditional computer systems [6].

2   Previous Work

ICS Honeypots are harder to build than other kinds of honeypots because they must simulate physical processes as well as standard network protocols. Nonetheless, several ICS honeypots have emerged. Conpot is the best-known (http://conpot.org), a low-interaction ICS honeypot which simulates basic ICS services including the EtherNet/IP, CIP, HTTP, S7, IPMI, SNMP, CACnet, and Modbus protocols. The GridPot honeypot, built as service of Conpot, simulates an ICS for an electric grid using the grid simulator GridLab-D [7] that provides realistic power-plant data. We developed an enhanced version of GridPot for the experiments reported here.

Other projects have built useful honeypots for ICS systems [8]. One project ran a large-scale low-interaction ICS honeypot using the Amazon EC2 cloud service for 28 days [9]. It serviced several protocols including Modbus, DNP3, ICCP, IEC 104, SNMP (most of the traffic), TFTP, and XMPP. Another project mimicked an electrical provider's network with an information-technology environment including a data network, an operational-technology environment operating physical processes and machinery, and a controller interface [10]. Soon after its launch, it was subject to ransomware attacks. A commercially available network-appliance honeypot was designed to model electrical-power grids [11] using GridPot. Another project tried to build an ICS honeypot better based on physics, and demonstrated applications to a heating-ventilation system and a water-treatment plant [12]. Two other honeypot prototypes for ICSs are described in [13] and [14]. We have previously studied several other kinds of honeypots [15, 16].

3   Experiments

Our experiments started with an earlier version of GridPot, and debugged and improved it. Client access to the GridPot came in through Conpot on ports 80 and 2404. Changes to the simulated data were sent to a GridLab-D simulation [17] and status changes were returned to the client through Conpot.

Our experiments had five phases. Phases 1 and 2 were run on a network connection through an Internet service provider and outside our organization's firewall. Phases 3, 4, and 5 were run with the cloud service provider DigitalOcean. More details of the implementations are in [18] and [19].

3.1   Design of Phases 1–2

Our design instantiated the default Conpot template to handle the IEC 60870-5-104 protocol ("IEC 104" for short), and it communicated with GridPot and the GridLab-D simulator. IEC 61850 was used in the original GridPot implementation, and is becoming an industry standard for electrical-substation automation. However, IEC 61850 is substantially more complex than other electric-grid ICS protocols because it specifies a data model, and IEC 61850 products must provide logical descriptions of every device used in the power grid. The IEC 104 protocol is much less prescriptive than IEC 61850, eliminating configuration of the data model as a possible source of error. Since Conpot and our SCADA application both supported full IEC 104 messaging, we chose the IEC 104 protocol for simulating control messaging in our honeypot.

To use IEC 104 with GridPot, a user sends an IEC 104 message to Conpot's IEC 104 server, which calls a GridPot simulator method to pass the command to the associated GridPot information object. It is then passed to the GridLAB-D simulation which makes any necessary changes and reports back to GridPot, which responds to the user via IEC 104. When Conpot queries a variable value, the GridPot simulator tells that GridPot object to poll its variables and report it.

Our Phase 2 enhanced GridPot with a SCADA (supervisory control) station with a graphical user interface to control the honeypot [20] in the hope of encouraging more direct user interaction with the simulated grid. Users could connect to a virtual machine on a Windows operating system through Microsoft's Remote Desktop Protocol and communicate with a backend Linux virtual machine running the Phase 1 GridPot. This meant that packets on the Windows host's inward-facing virtual network interface controller will show only IEC 104 traffic of the types expected between a real SCADA control station and its associated IEC 104 server.

3.2   Changes to the Conpot and GridPot Servers

We gave Conpot ability to distinguish two kinds of IEC 104 information-object variables in the server's template. GridPot variables mapped to a value in the GridLAB-D simulation, and change whenever they change in the simulation; Python variables mapped to variables in a GridPot object and can only change from user commands. For example, a GridPot switch might have an "enable" flag that must be set to True before the switch can be changed.

We also modified the Conpot IEC 104 server to handle additional commands not in the original GridPot that would increase the realism of the grid simulation. We changed the handling of IEC 104 type-45 (set single-point Boolean), type-46 (set double-point Boolean), type-49 (set scaled value), and type-100 (general interrogation) commands to let them respond without requiring a SCADA application, and added support for type-63 (set floating-point value with time tag) commands for when a user entered a floating-point number in the operator interface.

In Phase 2, we modified the Conpot IEC 104 server to send data updates to a remote user without prompting. To aid handling of updates, we changed the IEC 104 server to provide a pointer to its update handler for the GridPot simulator using the Conpot data bus, and accept a pointer to the simulator's update handler in return. While this could cause a crash if the server tries to retrieve a nonexistent pointer, testing did not see that.

We changed how the IEC 104 server communicated with our SCADA application. IEC 104 specifies that to change a variable, the remote station must send an "activation" command, to which the server responds with an "activation confirmation" message. The user then commands the server to change the variable and send back both an I-frame with the updated value and an "activation termination" message for the variable. Our SCADA application expects the updated value to come after the termination message and will not listen for it until the termination message has been received, although the documentation was unclear.

We expanded the functionality of the GL_obj class in GridPot to implement methods for object variables not in the GridLAB-D simulation, like the "Python" variables in the Conpot IEC 104 server. We also provided methods for converting between the strings GridLAB-D uses to store data and IEC 104 values which include complex numbers. These changes substantially reduced the amount of device-specific code required for each subclass of GL_obj, as for instance, we could add basic power meters as GridPot devices by instantiating the base GL_obj class without any device-specific code.

We did not simulate the full IEEE 13-Node Model because testing showed it slowed GridPot's operation unacceptably, so we simulated a voltage regulator, a switch, and seven power meters for residential customers. We created two main displays within our SCADA application: a report on the state of all simulated devices in the system, and details of a selected simulated device together with options that allowed users to change the device's parameters. To lure attackers, the SCADA host provided access to the system by the remote-desktop Guest account, which by default lacks a password and is a popular target for attack [21]. Providing an account without a password required changing the local machine's security policy and the Windows firewall rules.

3.3 Phases 1–2 Deployment

Phases 1 and 2 ran the honeypot as a virtual machine under VirtualBox 5.2.22 with two virtual processors, 10 GB of memory, and 50 GB of storage in a virtual disk. The physical machine had a bridged adapter, and used the same hardware (MAC) address as that of the host machine's network-interface controller through which the virtual machine was bridged. The controller's IP address, subnet, and gateway were manually configured to match the host machine's configuration. This allowed Conpot to record the remote IP addresses connecting to its open ports.

In Phase 2 another virtual network-interface controller was attached to the machine to support the graphical user interface. It was itself attached to a VirtualBox internal network equipped with a DHCP server to communicate with the Windows virtual machine hosting the SCADA application. The latter had four virtual processors, 6052 megabytes of memory, and 40 GB of storage. It had two virtual network interfaces; one was connected to the same VirtualBox internal network that the Phase 1 GridPot virtual machine used to communicate with the GridPot virtual machine, and the other was a bridged adapter with its MAC address set to the same address as the host machine's Internet-facing controller. Since the Windows host could not respond to HTTP traffic, this controller forwarded incoming traffic addressed for TCP port 80 to the internal network-facing controller. This allowed the GridPot HTTP server to respond to these requests.

We collected packet data with Wireshark to compile statistics on the honeypot traffic. Since the Microsoft Remote Desktop Protocol is encrypted, we used the security tool Mimikatz to extract the Windows host's encryption keys, allowing Wireshark to collect traffic in the clear. To restrict packet-capture files to a manageable size, we only recorded packets for the services we studied, HTTP (port 80) and IEC 104 (port 2404). Much IEC 104 traffic was incorrectly formatted and lacked the required starting byte of 0x68. We also collected data from Conpot and Windows logs; the latter were the Windows System, Windows Security, Windows Firewall, Microsoft-Windows-Terminal Services-LocalSessionManager, Microsoft-Windows-Terminal Services-RemoteConnectionManager, Microsoft-Windows-User Profile Service- Operational, Microsoft-Windows-WindowsDefender-Operational, and the Microsoft-Windows-User Profile Service-Operational. Since logs were continuously written to disk, they provided information for periods in which we lost packet captures, though some logs did not record the correct originating IP addresses. We also had trouble with Conpot logs because in some crashes the logger would stop and suddenly output a long list of events at the time it was stopped.

3.4   Phases 3–5 Deployment

We wanted to test whether cloud deployment of our honeypots would see different user behavior since it would involve an easier installation than traditional deployments. Our Phases 3–5 used the cloud provider DigitalOcean. We installed the same honeypot setup with Conpot and GridPot used in Phase 1, but without the SCADA interface because of the problems described in Sect. 4.2. Phase 3 was installed in a virtual machine for easier portability, and Phases 4 and 5 used virtualization methods provided by Digital Ocean. We got baseline data in Phase 3, then further obfuscated the honeypot and ran it at two sites, one in California (Phase 4) and one in Asia (Phase 5), to test regional differences in attacks. Phases 4 and 5 used additional obfuscations (the same for both) beyond Phase 3 in the names used within the honeypot.

Packet data was collected with the software Wireshark and Tshark. Packet capturing was configured to roll over into a new file every 25 MB, but this created difficulty for transfer to repositories, so we reduced it to 10 MB. Conpot logs were also collected that recorded honeypot-layer activity, and they were useful as backup when other data was unavailable due to crashes.

In Phase 4, the virtual desktop environment initially ran on top of the Ubuntu Linux operating system, but the environment stopped working and we had to use the default terminal console provided by the operating system. When invoked from the console, the bash command to open additional terminal instances did not work, so we executed our programs as background processes. This also required us to switch our packet-capture software to Tshark as Wireshark required a desktop environment for installation. Since background processes are suspended when a secure-shell session ends, Conpot, GridPot, and Tshark were initialized from the "droplet" console on DigitalOcean, and the secure shell was only used for system monitoring, data collection, and data transfer.

3.5   Data Analysis Methods

We wrote a Python program to parse PCAP input files and Conpot log files and record the timestamp, remote IP address, target TCP port, and HTTP method or IEC 104 frame type. For IEC 104 packets, the program also analyzed whether they contained valid frames, defined as TCP segments directed to port 2404 with a payload starting with a byte value of 0×68 and a second byte correctly specifying the length of the message; many packets directed at port 2404 were in HTTP format although that could not possibly work. The remote IP address for each request was also queried in GeoLite2, a free IP geolocation database. The resulting data was then saved into a Pandas DataFrame for analysis.

To determine which IP addresses contacted the honeypot more than once, we wrote another program. A session was defined as all packets exchanged by a socket pair on a day. To more closely inspect IEC 104 traffic, we wrote a program using Scapy to generate IEC 104 packets from Conpot log data for periods in which we lacked PCAP data. We could do this because Conpot logs for IEC 104 messages provide the sending and receiving sockets as well as the payloads, but not for HTTP traffic because Conpot does not capture the full payload of HTTP messages.

4   Results

4.1   Network Scanning

To test whether our honeypots were easy to detect, we used network scanning tools. The Shodan Honeyscore evaluation of our live Phase 1 GridPot said it found a 0% probability of our Phase 1 GridPot being a honeypot. A score of 1.0 (or 100% probability of being a honeypot) was received by the previous version of GridPot [6], indicating that our version is substantially better at evading a commonly used honeypot-detection tool. We also used the Shodan application programming interface to request information on our deployed GridPot. Shodan did not identify it as a honeypot regardless of whether the its history was considered. We conclude that our version of GridPot is better at evading this Shodan-based detection mechanism. We also checked our Phases 3–5 honeypots with the Shodan tool Honeyscore and got a score a score of 0.0 stating they were definitely not honeypots; however, Honeyscore has not been updated recently.

4.2   Compromises in Phase 2

Phase 2 was particularly interesting because our honeypot was compromised twice through the graphical user interface. The first run began on 5/26/20 and indicated compromise on 6/7 when we discovered a process "XMRig Miner" using all our processor. Further investigation found that packet capture had stopped, and a third-party software "Crack_by_NERO" had been installed and locked with a password on the Guest desktop. We found a new folder on the Guest desktop called "xmrig-5.1.1" which appeared to be a crypto-currency miner, as well as a desktop locker executable and a binary file "SpoolerComp.exe". The password locker achieved persistence by adding itself to the Windows Registry to run automatically on login.

We concluded that malware was installed in the Desktop folder because a Guest user had write access to it. We therefore disabled Guest's write access to Desktop and other folders in their home directory, as well as setting our SCADA software to start at login to make it more apparent that the system was an ICS. A second run was started on 6/17. When we checked the honeypot on 6/29, packet capture was no longer running, the Windows preinstalled Administrator account was visible despite it being made invisible at startup, and a new Administrator-level account called "Admin" had been created. A new executable "opera_portable_56.0.3051.36.exe" had been downloaded to the C:\Program Files directory, as well as several executables labeled as Opera installers.

We concluded that the system was penetrated at least twice on 6/25. The first used the Admin account and installed the infected opera.exe file, but was apparently thwarted by Windows Defender's anti-malware rules. This attack probably originated from the IP address 88.209.137.9. The second attack installed a different file in the same C:\ProgramData directory, which was again stopped by Windows Defender. Then something deactivated Windows Defender and proceeded with the exploit. Although log erasure means we cannot be certain whether these two attacks were related, the occurrence of the same type of malware in the same unauthorized C:\ProgramData directory suggests it.

4.3   Overall Comparison of the Phases

Table 1 shows statistics on the phases so far (as Phases 4 and 5 are still running); Phase 2 provided insufficient data to be useful since much was erased. Sessions were defined as all packets exchanged by a site (socket) pair on a day.

At first significantly more traffic occurred on the cloud sites in Phase 3 than on the standalone non-cloud sites, indicating that attackers were not discouraged by the cloud implementation. The cosine similarity between the Phase 1 and Phase 3 protocol distributions was 0.998, so the protocol variation was not significantly different. We did see significant differences in the traffic volume between the U.S. and Asia deployments (Phases 4 and 5), which suggest benefits of international cloud services for honeypots. Overall, we confirmed that cloud honeypots are feasible and effective for collecting intelligence on new cyberattacks on industrial control systems.

Table 2 compares the number of sessions and IP addresses between Phases 1 and 3. This suggests that addresses doing single sessions did relatively simple port scans, while others did more detailed investigation of the port. The mean of the weekly fraction

Table 1. Statistics on sessions during the experiments.

Phase	Phase 1	Phase 3	Phase 4	Phase 5
Days	157	17	18	18
Number of addresses that had a single HTTP session	2747	506	640	3740
Number of addresses that had multiple HTTP sessions and no ICS sessions	548	118	127	2464
Number of addresses that had a single ICS session	53	4	9	8
Number of addresses that had multiple ICS sessions and no HTTP sessions	32	1	3	3
Number of addresses that had both HTTP and ICS sessions	20	31	3	3
Total number of sessions	7121	1811	1515	21474

of traffic using HTTP over 22 weeks in Phase 1 was 0.934 with a standard deviation of 0.128, so traffic was primarily HTTP.

Table 2. Number of sessions and IP addresses between Phases 1 and 3.

Phase	Protocol	Number of sessions	Mean	Standard deviation	Number of addresses
1	HTTP	One	1.97	33.67	2235
1	HTTP	Multiple	38.35	169.0	573
3	HTTP	One	1.37	2.96	399
3	HTTP	Multiple	44.9	271.7	139
1	IEC 104	One	1.70	0.93	40
1	IEC 104	Multiple	15.42	13.15	38
3	IEC 104	One	2.63	1.27	16.4
3	IEC 104	Multiple	7.68	5.50	19.4

Table 3 further breaks down the IEC 104 traffic seen in all phases. Bearing in mind that Phase 1 ran for a longer period of time than the subsequent phases, Phase 3 actually had the highest rate of IEC 104 activity. The fraction of malformed IEC 104 packets varied significantly from week to week, with a mean over 20 weeks in Phase 1 of 0.505, standard deviation of 0.331, minimum of 0.0, and maximum of 1.0. From this we conclude that valid packets only occur in occasional campaigns and are not broadcast routinely, unlike the HTTP packets.

Figure 1 shows that inter-session times for Phase 1 differed significantly between HTTP (on the left) and IEC 104 (on the right). (Sessions were defined as all packets between the same pair of addresses on a day.). Rapid sequences of packets from a single

Table 3. Types of IEC 104 packets observed.

Phase	U-format frames	I-format frames	Error frames	Total IEC 104 traffic
Phase 1	78	19	557	654
Phase 3	13	4	171	188
Phase 4	11	4	13	28
Phase 5	8	2	14	24

IP address might indicate scanning activity. ICS traffic had short gaps, but HTTP traffic had a wider range with a local peak between 10,000 and 100,000 s.

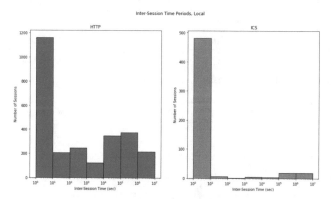

Fig. 1. Inter-packet times for Phase 1

Table 4 shows that the reported countries of origin of the sessions in the experiments were reasonably consistent.

Figure 2 shows a sample of the traffic rates for Phase 1, and Fig. 3 shows them for Phase 3. These plot the logarithm of the number of incoming requests per day for the honeypots. HTTP traffic is in blue; "IEC 104 traffic" in orange means all traffic using IEC 104's TCP port 2404; and "IEC 104 messages" in green means only traffic with correctly formatted IEC 104 data. Figure 2 shows more data because it ran longer than Phase 3. Phase 1 saw HTTP on most days; Phase 3 saw continuous HTTP and IEC 104 probing. However, both phases saw gaps in traffic with valid IEC 104 messages because of its low overall frequency; most IEC 104 traffic was invalid, which most likely means basic port scanning.

Table 4. Percentages of claimed countries of origin.

	Phase 1		Phase 3		Phase 4		Phase 5	
	Country	*%*	*Country*	*%*	*Country*	*%*	*Country*	*%*
1	China	22.5	United States	40.7	United States	34.6	United States	40.9
2	United States	16.5	Russia	16	China	16.5	Singapore	13.5
3	Russia	12.3	China	9.1	Russia	10.7	Canada	4.9
4	Taiwan	4.7	Germany	6.4	Netherlands	6.4	Nether-lands	4.6
5	Brazil	3.8	Nether-lands	3.2	United Kingdom	5.4	France	4.5
6	Nether-lands	3.6	Brazil	2.6	India	3	Germany	3.4
7	Germ-any	3	Romania	2.4	Brazil	2.3	China	2.3
8	Japan	1.8	Seychelles	2.2	Germany	1.9	Italy	2.1
9	India	1.7	India	1.8	France	1.7	Sweden	2.1
10	Canada	1.6	Switzer-land	1.2	Romania	1.7	United Kingdom	2.0
11	Other	28.5	Other	14.5	Other	15.9	Other	19.6

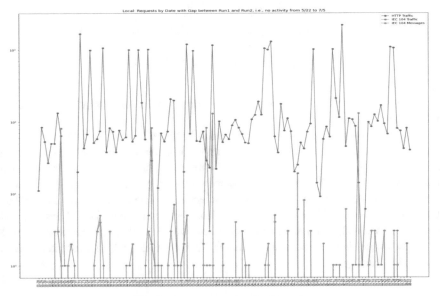

Fig. 2. Requests per day for Phase 1.

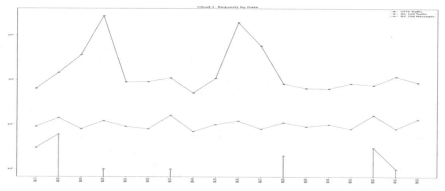

Fig. 3. Requests per day for Phase 3.

5 Conclusions and Future Work

We saw similar HTTP traffic in the cloud environment of Phases 3–5 than in the physical environment of Phases 1–2, but less IEC 104 traffic in all phases. This could be due to the rarity of industrial control systems in the cloud and disinterest by adversaries in attacking them due to lack of familiarity with them, but this could change as ICS systems become more common.

We added more obfuscations to Phases 4 and 5 to make the GridPot less obvious. Traffic decreased, but it may have only been due to the decreased novelty of the configuration, since a less obvious honeypot should be more desirable to attack. Phase 5 showed significantly more traffic in an Asia-based deployment rather than a US-based deployment, which again may have been due to a novelty effect. However, much data appeared to be due to an previous DNS poisoning attack on the service which may have increased the traffic. This can be an issue when using a cloud service that recycles Internet addresses.

Phases 4 and 5 are still collecting data, and Phase 2 has been resumed on a Linux machine using a Windows Applications Programming translator to avoid the vulnerabilities of the Microsoft environment. We anticipate having more data to see if the trends continue.

Currently we are exploring T-Pot, a multi-honeypot platform, to see if providing better responses to HTTP requests and responses to additional protocols will increase effectiveness of a honeypot. We also identified a section of code in the GridPot implementation that could be modified to make it especially less obvious. Having more detailed simulation capabilities for industrial control systems increased cyberattacker interest and the amount of activity on our honeypots, so those simulation features should be enhanced. As for data analysis, our data finds new cyberattacks although that was not our focus so far. Finally, since we observed regional differences, more honeypots could be set up all over the world using cloud services. Being able to compare attacks across many sites should make it easier to see past the randomization of many attacks.

References

1. Stouffer, K., Pillitteri, V., Lightman, S., Abrams, M., Hahn, A.: Guide to industrial control systems (ICS) security. In: National Institute of Standards and Technology, Gaithersburg, MD, US, NIST SP 800-82 Revision 2 (2015)
2. Hawk, C., Kaushiva, A.: Cybersecurity and the smarter grid. Electr. J. **27**(8), 84–95 (2014)
3. Blume, S.: System overview, terminology, and basic concepts. In: Electrical Power System Basics for the Nonelectrical Professional, pp. 1–12. Wiley, Hoboken (2007)
4. Greenberg, A: Crash override: the malware that took down a power grid. Wired (2017)
5. Atadika, M., Burke, K., Rowe, N.: Critical risk management practices to mitigate cloud migration misconfigurations. In: Proceedings of International Conference on Computational Science and Computational Intelligence, Las Vegas, NV, United States (2019)
6. Rowe, N., Nguyen, T., Kendrick, M., Rucker, Z., Hyun, D., Brown, J.: Creating effective industrial-control-systems honeypots. In: Proceedings of Hawaii International Conference on Systems Sciences, Wailea, HI, USA (2020)
7. Redwood, W.: Cyber physical system vulnerability research. Ph.D. dissertation, Florida State University (2016)
8. Dalamagkas, C., et al.: A survey on honeypots, honeynets and their applications on the smart grid. In: Proceedings of the 2019 IEEE Conference on Network Softwarization, Paris, France (2019)
9. Serbanescu, A., Obermeier, S., Yu, D.-Y.: ICS threat analysis using a large-scale honeynet. In: Proceedings of 3rd International Symposium for ICS and SCADA Cyber Security Research, pp. 20–30 (2015)
10. Barak, I.: Cybereason's newest honeypot shows how multistage ransomware attacks should have critical infrastructure providers on high alert. J. Cyber Policy (2020)
11. Quantalytics, Q GridPot. www.quantalytics.com/q-GridPot/. Accessed 18 Jan 2019
12. Litchfield, S.: Honeyphy: a physics-aware CPS honeypot framework. Master's thesis, Georgia Institute of Technology, Atlanta, GA, US (2017)
13. Buza, D., Juhász, F., Miru, G., Félegyházi, M., Holczer, T.: CryPLH: protecting smart energy systems from targeted attacks with a PLC honeypot. In: Proceedings of International Workshop on Smart Grid Security, pp. 181–192 (2014)
14. Antonioli, D., Agrawal, A., Tippenhauer, N.: Towards high-interaction virtual ICS honeypots-in-a-box. In: Proceedings of 2nd ACM Workshop on Cyber-Physical Systems Security and Privacy, pp. 13–22 (2016)
15. Rowe, N., Rrushi, J.: Introduction to Cyberdeception. Springer, Cham (2016). https://doi.org/10.1007/978-3-319-41187-3
16. Rowe, N.C.: Honeypot deception tactics. In: Al-Shaer, E., Wei, J., Hamlen, K.W., Wang, C. (eds.) Autonomous Cyber Deception, pp. 35–45. Springer, Cham (2019). https://doi.org/10.1007/978-3-030-02110-8_3
17. GridLAB-D Project: GridLAB-D Github page. https://github.com/gridlab-d/gridlab-d. Accessed 09 Aug 2020
18. Dougherty, J.: Evasion of honeypot detection mechanisms through improved interactivity of ICS-based systems. Master's thesis, U.S. Naval Postgraduate School (2020)
19. Bieker, M., Pilkington, D.: Deploying an ICS honeypot in a cloud computing environment and comparatively analyzing results against physical network deployment. Master's thesis, U.S. Naval Postgraduate School (2020)
20. Paganini, A.: IndigoSCADA User Manual, rev. 334. http://www.enscada.com/a7khg9/Indigo SCADA_user_manual.pdf. Accessed 13 Aug 2019
21. Boddy, M., Jones, B., Stockley, M.: RDP exposed - the threat that's already at your door. In: Sophos White paper. Sophos, Inc. (2019)

Risks of Electric Vehicle Supply Equipment Integration Within Building Energy Management System Environments: A Look at Remote Attack Surface and Implications

Roland Varriale[1]([✉]), Ryan Crawford[2], and Michael Jaynes[1]

[1] Argonne National Laboratory, Lemont, IL 60439, USA
{rvarriale,mjaynes}@anl.gov
[2] Illinois Institute of Technology, Chicago, IL 60616, USA
rcrawford@hawk.iit.edu

Abstract. The rapid development of Electric Vehicle Supply Equipment (EVSE) and incorporation within more traditional industrial control systems, such as Building Management Systems (BMS) and Building Energy Management Systems (BEMS) can lead to undesirable consequences. In a desire to gain functionality, separated components are often integrated and accessible through less secure means. Furthermore, the architecture and operational environment of some devices may lead to unintended consequences based on either physical or logical co-location. To perform analysis of the remote attack surface we used publicly available tools and data sets such as Shodan [1], nmap [2], and exploit-db's searchsploit tool [3].

We will also discuss evidence of possible remote attack surface weakness across the sector, as a whole, and what the implications of weaknesses within our threat model may permit within the operating environment.

Keywords: Electric Vehicle Supply Equipment · Building Energy Management system · Cyber security · Electrification · Open source intelligence · Vulnerability analysis · Threat model · Attack graph

1 Introduction

EVSE contain charging infrastructure that delivers power to an electric vehicle. This infrastructure provides charging capabilities at a variety of different standards and power levels. Level 1 and Level 2 chargers use alternating current and operate at 120 V and 240 V, respectively. More powerful Level 3, so-called DC-Fast Chargers, chargers operate using direct current and can operate at much higher voltages. Current research within EVSE looks at applications of Megawatt+ zcharging which allows for rapid charging of medium to heavy duty vehicles for use within logistics or delivery [4]. As EVSE technology and capability advances, so does the complexity within its implementation and integration.

© The Author(s), under exclusive license to Springer Nature Switzerland AG 2022
K.-K. R. Choo et al. (Eds.): NCS 2021, LNNS 310, pp. 163–173, 2022.
https://doi.org/10.1007/978-3-030-84614-5_13

This research explores the attack surface that could allow an outsider to gain access to the charging infrastructure and, subsequently, attached systems such as building energy management systems (BEMS), building automation systems (BAS), or building management systems (BMS) in order to attack the system's confidentiality, integrity, or availability.

Currently, EVSE are required to perform a myriad of functionality including housing payment processing, information technology (IT) networking, enterprise integration, and power consumption and management. EVSE also act as a conduit between manufacturer IT systems and client IT systems and can bridge the networks if not separated properly. This separation becomes even more important when integrating EVSE within systems where industrial control systems (ICS) are present, such as in BMS or BEMS. Logical, or trust, boundaries such as firewall rules aim to restrict device to device communication and restrict unwanted communication between devices.

1.1 Technical Background

Certain protocols used by the EVSE, such as Open Charge Alliance's Open Chargepoint Protocol (OCPP) [5] and the ISO 15118 specification family [6], are publicly known and available to research. OCPP, specifically, does not include public key infrastructure encryption until version 1.8, and the current version is 2.1, yet this functionality has not increased the adoption of the updated standard due to the need to rework previous systems. Wang et al. discuss safety protection inherent to different types of attacks against EVSE, including cyber security attacks [7]. Although there is a small section pertaining to potential economic losses due to availability disruption, it introduces the topic within the context of EVSE safety. Alacaraz et al. discuss threats associated with OCPP at v1.8 and below [8]. This lack of PKI provides one insight into the lag of security adoption within the EVSE space which should raise security considerations for their adoption within sensitive or real time networks such as BEMS. Furthermore, Falk and Fries expound upon the types of EVSE communication and possible attacks against system confidentiality, integrity, and availability [9].

Pratt and Carroll [10] provide an in-depth look into various threats surrounding EVSE and possible attack vectors. Their work describes the interplay between EVSE, personal electric vehicles (PEVs), and the standards that support them. This web of heterogeneous infrastructure, protocols, and standards provides a foundation for a more detailed inspection of the interplay between these systems.

A specific area of interest for this research lies within the operational environment created by integrating a newer technology with basic security practices (EVSE) within a real time (BEMS) environment. This research identified possible security threats by analyzing services and protocols present within these networks and concerns which arise when co-located on the same network.

Figure 1 depicts several logical layers between critical devices and the Internet, namely through the control system or the telephony firewalls. According to this model there are a minimum of two logical devices traversed in order to access devices within this zone. Our research has shown conflation of several zones into one where some EVSE host webservers as well as provide internal network access to enterprise networks and/or

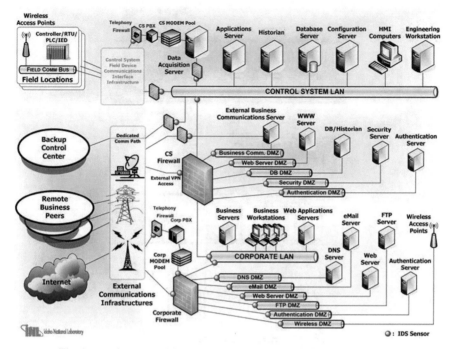

Fig. 1. A reference architecture based on separation of functionality [11]

BEMS/BMS. This architecture, supplied by the industrial control system computer emergency readiness team (ICS-CERT), provides a reference used by mature organizations to separate devices and communications based on purpose and security to promote secure operations. Shirking the tiered protections that several reference architectures lay out may allow attackers to pivot from an external to internal network without proper authentication. For example, a weakness in the web server such as a reused credential, common credential or known vulnerability can allow an attacker access to that device which then can perform additional reconnaissance or move laterally across the internal network. This presents a nominal problem since EVSEs can be easily discoverable through Google or Shodan searches when they are deployed on traditional networks. EVSE manufacturers often require cellular communication; however, we identified many devices present on the open Internet (Fig. 2).

The Purdue Enterprise Reference Architecture notes the separation of functionality into logical groups and informs the reference architecture provided by ICS-CERT in Fig. 1. This supports proper software and security engineering practices by establishing both separation of concerns and duties. ICS-CERT contains a repository of additional material[1] which explains the necessity of this separation and scrutiny in great detail. Many installments follow this separation initially; however, security exists as a continuum and degrades over time without constant maintenance. Likewise, some security controls do not properly address the risk associated with a component or system. This

---

[1] https://us-cert.cisa.gov/ics/Recommended-Practices.

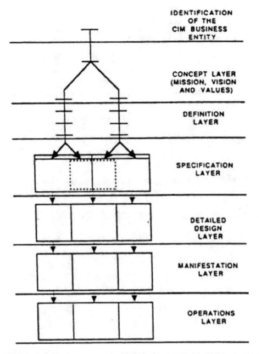

Fig. 2. The Purdue Enterprise Reference Architecture (PERA) [12], © 1993 International Federation of Automatic Control. Reproduced with permission from IFAC Proceedings Volumes, 26/2.

paper will not focus on a particular manufacturer or implementation; however, we investigate some common deficiencies demonstrated by the variety of systems we interacted with.

By not properly separating the remote access vectors with the operational technology (OT) controls, the control system network lacks proper separation. This can permit malicious actors to execute undesirable commands. According to the PERA, separation between those areas should be scrutinized by an authentication mechanism.

During the course of this paper, we will highlight some of the security considerations within the integrated environments that result from energy systems created from different vendors. Cyber physical systems, typically containing both IT and OT components, have been studied at length; however, implementation of secure architectures becomes very difficult when integrating dissimilar components among many different vendors. Reference architectures, such as the Purdue Enterprise Reference Architecture, provide guidelines towards best practices that reverberate through national standards such as NIST SP 800-53 rev5 [13].

2   EVSE Vulnerability Analysis Methodology

Our EVSE analysis methodology stemmed from an approach taken in open-source intelligence research. We compiled a list of keywords related to EVSE manufacturers and cast a wide net using both Shodan and Google. Once a candidate host was identified by exhibiting relevant metadata, we noted any additional attributes, keywords, ports, and other relevant information and included them within our search base. Using this technique, we identified several EVSE from different manufacturers. The search terms ranged from broad, such as a charging manufacturer, or very specific such as model names or terminology extracted from company-provided fact sheets.

Subsequently we utilized enumeration tools, such as Sublist3r[2], to gain information and correlate the different hosts. This allowed us to identify groups of identifiers and services that we mapped to specific manufacturers. Enumeration tools can allow for rapid identification of publicly available records and sites related to the provided host. Sublist3r ingests a hostname and identifies subdomains present within query results on search engines.

Next, we inspected the hosts using the nmap tool to identify any high-numbered open ports, service versions, and other related metadata that would inform our vulnerability analysis. Nmap probes one or more hosts and leverages response statistics to identify host characteristics such as operating system, port status, and service versioning. Nmap also contains a scripting engine which allows operators to employ pre-built or custom-built scripts to leverage the nmap engine for vulnerability analysis, enumerations, and other purposes. During our research we used the vuln keyword to invoke vulnerability scripts related to the services that were identified.

Finally, we correlated the services running with possible vulnerabilities and exploits, which may allow for remote code execution, privilege escalation, or other undesirable outcomes. The two main references for this task were NIST's NVD [14] and Exploit-DB [15]. The former contains a generic list of vulnerabilities related to specific services identified by a common vulnerability enumeration (CVE) number; whereas the latter contains a list of vulnerable services along with proof-of-concept code that demonstrates a tested exploitation of that vulnerability.

3 EVSE Findings

Many observed vulnerabilities were a result of either technical debt—advancement of functionality without the employment of commensurate security controls to preserve the system's secure operation—or lack of software updates. This can offer a misalignment of security and functionality which provides an avenue of exploitation if an attacker identifies this weakness. Similarly, Internet facing devices must update software as soon as feasible since they offer an enticing avenue of potential, organizational network entry to an attacker. Notably, devices running a web server were observed which may provide a conduit for access between internal and external networks.

Our research identified a deviation in best practices between manufacturer installation instructions and installation in reality. Required manufacturer services often become a victim of reconfiguration during system deployment and integration, which weakens security measures provided by the manufacturer. We will note several of these discrepancies and their cyber security consequences. To preserve the anonymity of the EVSE

---

[2] https://github.com/aboul3la/Sublist3r.

Pivoting Between Systems to Access Desired Data Flows

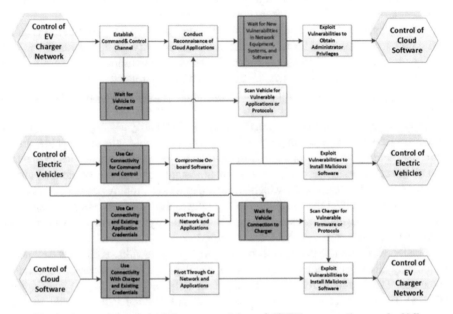

Fig. 3. An attack graph depicting traversal through EVSE connected networks [16]

manufacturers we will refer to general classes of services running on the EVSE and not specific ports or service versions which may uniquely identify the manufacturer. We note that all manufacturers did not exhibit these vulnerabilities; however, many did and will inform our analysis during this section.

The most prominent vulnerabilities lied within commonly used or weak credentials, lack of login timeout functions to dissuade brute-forcing attempts, running unnecessary services, and running outdated services. Many of these services offer unprivileged access to the EVSE; however, exploits could chain together, offering more privileged access to the EVSE system. This was often the case where one service offered remote access or remote code execution capabilities, but a second exploit was necessary to promote access to a more privileged user in order to examine memory contents or perform other interesting actions requiring super user or Administrator access.

4   Risks of Exploitation

The highlighted vulnerabilities within the remote attack surface may offer internal network access to an attacker who only has access to the single host. Possible exploitation results in access to the EVSE Network Presence, depicted in Fig. 3. An attack graph depicting traversal through EVSE connected networks [16]. Once entry has been gained, there are three interesting outcomes from a cyber-physical perspective. Firstly, the chargers can be randomly stopped from charging vehicles, which would result in Electric vehicles not being charged when needed, and it would also result in a loss of consumer confidence in chargers. Secondly, access could be used to install malicious implants to harvest payment information resulting in a loss of EVSE user financial. Lastly, access may be used to garner additional organizational information, gain access to sensitive networks, or exfiltrate intellectual property from the EVSE or their connected networks. We will omit the first outcome since the implications only affect availability and warrant no further examination. Some of the findings correlated with the prior findings of Lee *et al.* [17] regarding vulnerabilities within the implementation of ISO15118; however, we believe that some of the vulnerabilities highlighted within that work were also created during implementation and may not indicate weaknesses within the protocol itself (Fig. 4).

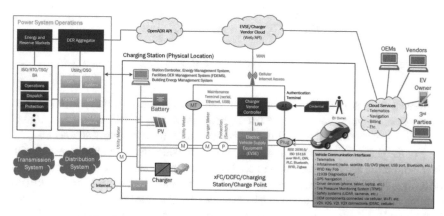

Fig. 4. An overview of EVSE ecosystem participants [18]

5   Expanding the EVSE Threat Model to Include BEMS/BMS

We extend the operational and attack surface to include both local and remote connections related to the integrated systems they incorporate within such as BMS and BEMS. Open-source BEMS documentation review uncovers the use of industrial control system protocols within these internal systems. As previously discussed, these protocols often communicate using insecure means and require defense-in-depth security approaches to properly secure their systems [18]. The graphic from NIST SP800-82 notes this in Fig. 5. Proper ICS network segmentation [19]. Figures 1, 5, and 6 offer insights into trust boundaries, reinforced by the logical rules within firewalls to separate communication based on need and functionality.

However, in practice the firewall depicted is often omitted in favor of increased communication between EVSE and BEMS. The authors posit that this occurs due to the need for many different systems to operate in a real-time environment and any changes

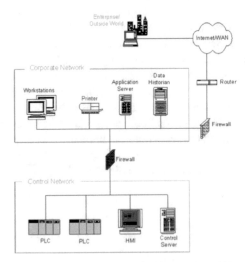

Fig. 5. Proper ICS network segmentation [20]

to firewall rules may disrupt that communication. This was observed on several hosts which were running protocols supporting both EVSE and BEMS operations, whereas these services should reside on different devices on different network segments.

For the purposes of our discussion, we do not consider the individual protocols since they exhibit the same properties, namely lacking authentication and authorization. This is to say, as a network device you can send commands to interact with the BEMS/BMS through non-sophisticated means such as media access control (MAC) spoofing or man-in-the-middle attacks. Some industrial protocols, such as Message Queue Telemetry Transport (MQTT), allow for authentication but it needs to be configured and only unauthenticated versions were observed in our experience [21]. Other observed protocols were (Building Automation Control network) BACnet and Modbus.

During this section we will extend Johnson and Anderson's attack graph [18] and identify potentially useful logical trust boundaries.

Extending this model, we can add nodes which incorporate traditional IT network communication and traverse a TCP to Serial converter or an ICS friendly Networking adapter. This highlights the need for logical separation of these devices which would reduce efficacy of traditional network-based attacks against ICS. Co-locating EVSE and BEMS/BMS devices allows for direct, unfettered communication between devices which further expands the reach of attackers through the highlighted remote attack surface. Hannan *et al.* note energy routers, within BEMS, as a potential data security risk [22]. This highlights a distinct area where a trust boundary, such as a firewall, must ensure desirable communication between devices (Fig. 7).

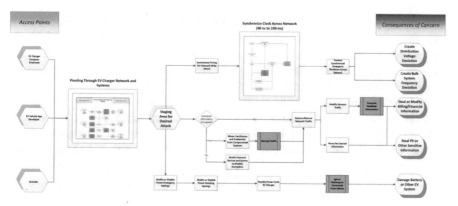

Fig. 6. Sample threat model of EVSE network with potential consequences [16]

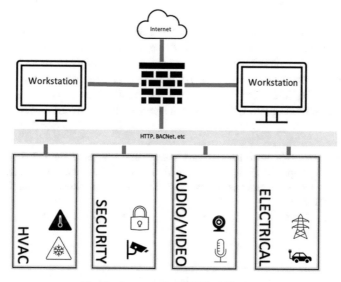

Fig. 7. An example BEMS architecture

6   Conclusions

This research points to improving and revisiting security plans around EVSE charging infrastructure integration within BEMS environment. Most notably, implementations where many systems may be integrated such as large public spaces, such as shopping areas. The need for interoperability may supersede the immediate need for security and build technical debt. This presents a formidable challenge when incorporating these systems within other systems since EVSE security is fragile and requires a robust, defense in depth approach to ensure compromise of EVSE does not compromise other organizational assets.

In the context of organizational IT security, both BEMS and EVSE represent third party systems requiring permissions and resource access on IT networks and systems. Current risk paradigms surrounding external dependencies or third-part inclusion leverage defense in depth solutions, least privilege access, and exercises assuming compromise of different components which you depend on [23, 24].

As functionality on EVSE and BEMS increase, lapses in securing or limiting third party devices and infrastructure may lead to supply chain vulnerabilities as demonstrated in the 2020 SolarWinds compromise [25]. Third party functionality and protocol integration increase the likelihood of a supply chain attack by increasing the attack surface. Without adherence to reference architectures advocating secure EVSE inclusion within BEMS, security issues may arise. Overall, this work motivates the need supplemental research to further understand how these architectures may work together more securely.

7 Future Research

Future research into this subject should investigate more possible vulnerabilities. This research mainly found possible vulnerabilities relating to EVSE. There are many manufacturers of electric vehicle supply equipment, so more of them should be explored. There are also many manufacturers of BEMS. By finding more documents that detail how specific devices integrate into the BEMS system, there can be a better understanding of how to securely integrate EVSE into a system with completely different industrial protocols.

Acknowledgement. This material is based upon work supported by Laboratory Directed Research and Development (LDRD) funding from Argonne National Laboratory, provided by the Director, Office of Science, of the U.S. Department of Energy under Contract No. DE-AC02-06CH11357.

References

1. Shodan. https://www.shodan.io. Accessed 14 Mar 2021
2. NMAP.ORG. https://nmap.org. Accessed 14 Mar 2021
3. Searchsploit - The Manual. https://www.exploit-db.com/searchsploit. Accessed 14 Mar 2021
4. Bohn, T.: Scalable electric submeter challenges for electric vehicle charging; low level AC to DC extreme fast charging for commercial vehicles. In: 2019 IEEE Transportation Electrification Conference and Expo (ITEC). IEEE (2019)
5. Open Charge Alliance. https://www.openchargealliance.org. Accessed 14 Mar 2021
6. International Standards Organization ISO 15118-1:2019 Road vehicles—vehicle to grid communication interface—Part 1: general information and use case definition. https://www.iso.org/standard/69113.html. Accessed 14 Mar 2021
7. Wang, B., et al.: Electrical safety considerations in large-scale electric vehicle charging stations. IEEE Trans. Ind. Appl. **55**(6), 6603–6612 (2019)
8. Alcaraz, C., Lopez, J., Wolthusen, S.: OCPP protocol: security threats and challenges. IEEE Trans. Smart Grid **8**(5), 2452–2459 (2017)
9. Falk, R., Fries, S.: Electric vehicle charging infrastructure security considerations and approaches. In: Proceedings of INTERNET, pp. 58–64 (2012)

10. Pratt, R.M., Carroll, T.E.: Vehicle charging infrastructure security. In: 2019 IEEE International Conference on Consumer Electronics (ICCE). IEEE (2019)
11. Secure Architecture Design, United States Computer Emergency Readiness Team. https://us-cert.cisa.gov/ics/Secure-Architecture-Design. Accessed 14 Mar 2021
12. Williams, T.J.: The Purdue enterprise reference architecture. Comput. Ind. **24**(2–3), 141–158 (1994)
13. Security and Privacy Controls for Information Systems and Organizations, National Institute for Standards and Technology, United States Department of Commerce. https://csrc.nist.gov/publications/detail/sp/800-53/rev-5/final. Sept 2020
14. National Vulnerability Database. https://nvd.nist.gov. Accessed 14 Mar 2021
15. Exploit Database. https://www.exploit-db.com. Accessed 14 Mar 2021
16. Anderson, B.R., Johnson, J.T.: Securing Vehicle Charging Infrastructure Against Cybersecurity Threats. No. SAND2020–0818C. Sandia National Lab. (SNL-NM), Albuquerque, NM (United States) (2020)
17. Lee, S., et al.: Study on analysis of security vulnerabilities and countermeasures in ISO/IEC 15118 based electric vehicle charging technology. In: 2014 International Conference on IT Convergence and Security (ICITCS). IEEE (2014)
18. Johnson, J.T.: Securing Vehicle Charging Infrastructure. No. SAND2019–4145PE. Sandia National Lab. (SNL-NM), Albuquerque, NM (United States) (2019)
19. Rhode, K.: VTO diagnostic security modules for electric vehicle to building integration, 07 June 2016. https://www.energy.gov/sites/prod/files/2016/06/f32/vs184_rohde_2016_p_web.pdf. Accessed 18 Dec 2020
20. https://nvlpubs.nist.gov/nistpubs/SpecialPublications/NIST.SP.800-82r2.pdf
21. Matabuena, D., et al.: Device for smart loads management in building energy management system. In: Proceedings of Seminario Anual de Automática, Electrón. Ind. Instrum. (2018)
22. Hannan, M.A., et al.: A review of internet of energy based building energy management systems: Issues and recommendations. IEEE Access **6**, 38997–39014 (2018)
23. Densham, B.: Three cyber-security strategies to mitigate the impact of a data breach. Netw. Secur. **2015**(1), 5–8 (2015)
24. Basiri, A., et al.: Chaos engineering. IEEE Softw. **33**(3), 35–41 (2016)
25. Ballard, B.: Microsoft says it has identified over 40 victims of SolarWinds hack, 18 Dec 2020. https://www.techradar.com/news/microsoft-says-it-has-identified-over-40-victims-of-solarwinds-hack. Accessed 18 Dec 2020

Author Index

K.-K. R. Choo et al. (Eds.): NCS 2021, LNNS 310, pp. 175, 2022.
https://doi.org/10.1007/978-3-030-84614-5